Sustainable Production, Life Cycle Engineering and Management

Series editors

Christoph Herrmann, Braunschweig, Germany
Sami Kara, School of Mechanical and Manufacturing Engineering, The University of New South Wales, Sydney, NSW, Australia

Modern production enables a high standard of living worldwide through products and services. Global responsibility requires a comprehensive integration of sustainable development fostered by new paradigms, innovative technologies, methods and tools as well as business models. Minimizing material and energy usage, adapting material and energy flows to better fit natural process capacities, and changing consumption behaviour are important aspects of future production. A life cycle perspective and an integrated economic, ecological and social evaluation are essential requirements in management and engineering. This series will focus on the issues and latest developments towards sustainability in production based on life cycle thinking. To submit a proposal or request further information, please use the PDF Proposal Form or contact directly: *Silvia Schilgerius, Publishing Editor (Silvia. Schilgerius@Springer.com)*

More information about this series at http://www.springer.com/series/10615

Liselotte Schebek · Christoph Herrmann
Felipe Cerdas
Editors

Progress in Life Cycle Assessment

 Springer

Editors
Liselotte Schebek
Material Flow Management
and Resource Economy
Technische Universität Darmstadt
Darmstadt, Hessen, Germany

Felipe Cerdas
Sustainable Manufacturing and Life
Cycle Engineering
Technische Universität Braunschweig
Braunschweig, Germany

Christoph Herrmann
Sustainable Manufacturing and Life
Cycle Engineering
Technische Universität Braunschweig
Braunschweig, Germany

ISSN 2194-0541   ISSN 2194-055X   (electronic)
Sustainable Production, Life Cycle Engineering and Management
ISBN 978-3-030-06390-0   ISBN 978-3-319-92237-9   (eBook)
https://doi.org/10.1007/978-3-319-92237-9

This Springer imprint is published by the registered company Springer Nature Switzerland AG
The registered company address is: Gewerbestrasse 11, 6330 Cham, Switzerland

Preface

The provision of products (goods and services) rises environmental challenges on a life cycle perspective, i.e. from cradle to grave. Life cycle assessment (LCA) is a system analysis methodology developed to ease the inventorization and evaluation of both the material and energy demanded as well as the emissions generated by product systems throughout their supply chain, usage and end-of-life stages.

While the methodology is fairly comprehensive, it still needs additional research and development to further increase robustness and reliability. The more we understand the 'planetary boundaries' the more we realize that LCA should become a methodology for everyone, allowing the evaluation of environmental impacts in different application areas.

As the interest on the methodology steadily increases, the number of open questions and possible directions for future research grows as well. Topics of discussion such as issues of uncertainty, variability, regionalization, accessibility, homogenization of inventory modelling approaches, interpretation and, last but not least, data representativeness, quality and modelling, are persistently seen as possible methodological pitfalls within the LCA research community. In this regard, discussion platforms to enable an exchange between different research groups and the dissemination of scientific findings and results are essential drivers of innovation. The *Ökobilanzwerkstatt* (LCA workshop) intends to contribute to this objective. The present work, Progress in Life Cycle Assessment, summarizes the findings and scientific results of the latest research activities in the field of LCA presented at the 13th Ökobilanzwerkstatt jointly organized by the Technische Universität Darmstadt and the Technische Universität Braunschweig.

Darmstadt, Germany
Braunschweig, Germany
Braunschweig, Germany

Prof. Dr. rer. nat. Liselotte Schebek
Prof. Dr.-Ing. Christoph Herrmann
Felipe Cerdas, M.Sc.

Contents

Part I
Introduction

State of the Art and Future Developments in LCA

Tobias Viere

Abstract This essay serves as a preface to the contributions from young LCA researchers that participated in the 13th Ökobilanzwerkstatt ("LCA workshop") at Braunschweig's Technical University. It explains the basic idea of LCA and briefly explores some of the most important future developments and related research demands.

Keywords LCA · Ökobilanzwerkstatt

1 Introduction

Resource scarcities, environmental pollution and the need for renewable energies have been in the centre of interest for a few decades already. As early as in the 1970s researchers and decision makers were interested in the resource use and environmental damage implications of particular products and packaging options. This has been the starting point for the development of the Life Cycle Assessment (LCA) methodology, which is nowadays the most established and widely used method to assess the environmental impacts of products, services and technologies. This essay explains the basic idea of LCA and its current state in business practice. It explores some of the most important future developments and related research demands. The essay is part of a book that summarizes contributions from young LCA researchers that participated in the 13th Ökobilanzwerkstatt ("LCA workshop") at Technische Universität Braunschweig, Germany, in September 2017. Hence, the essay finally elaborates the role of the Ökobilanzwerkstatt for progressing LCA research and practice.

Most useful references to LCA and its developments can be found in textbooks (e.g. Hauschild et al. 2018), standards and guidelines (e.g. ISO 14040 and 14044

T. Viere (✉)
Institute for Industrial Ecology, Pforzheim University,
Tiefenbronner Str. 65, 75175 Pforzheim, Germany
e-mail: tobias.viere@hs-pforzheim.de

© Springer Nature Switzerland AG 2019
L. Schebek et al. (eds.), *Progress in Life Cycle Assessment*, Sustainable
Production, Life Cycle Engineering and Management,
https://doi.org/10.1007/978-3-319-92237-9_1

3

and EC-JRC 2010), and journals (International Journal of Life Cycle Assessment, Journal of Cleaner Production, and Journal of Industrial Ecology, to name a few).

2 LCA in a Nutshell

Life Cycle Assessment (LCA) is used to assess environmental impacts of product and service systems over the whole life cycle from raw material extraction to end of life. LCA is used for decision making on various levels and within different functions and organizations including product comparisons and technology evaluations on corporate levels up to political and macro-scale studies on the effects of environmental policies. The common LCA approach following ISO 14040 and 14044 distinguishes four essential process steps required in all types of studies: goal and scope, life cycle inventory, life cycle impact assessment, and interpretation, all of which are elaborated in an iterative manner.

Goal and scope defines the framework of a study including spatial, temporal and technical requirements and system boundaries. Furthermore, it defines the study's functional unit, which describes the product or service system's benefit. All products or technologies under comparison and their impacts as well as different scenarios of the same product systems are scaled to the same functional unit in order to ensure comparability. The Life Cycle Inventory (LCI) records all material and energy flows (raw materials, emissions, intermediates etc.) that flow into or out of the system under assessment. While some LCI data for very important and focal steps of the life cycle is measured and gathered directly, much data for life cycle processes further up or down the chain is derived from generic LCI databases. Within Life Cycle Impact Assessment (LCIA) the LCI records are assessed according to their contribution to certain environmental impact categories such as global warming, eutrophication, resource depletion, or land use change. The Interpretation step within LCA evaluates the results further and discloses uncertainties and insufficiencies of previous steps in order to improve the overall assessment.

3 LCA Development

Since its early days more than 40 years ago, LCA has constantly developed and diversified. For instance, the well-known carbon footprint follows LCA procedures with its LCIA focusing on one particular environmental impact only, namely the global warming potential. Instead of product and services, corporate carbon footprints and organization's environmental footprints extent the idea of LCA to organizations and their performance over time. Besides environmental impacts, LCA is also capable of considering economic and social aspects in life cycle cost assessments or Social LCA respectively. This diversification and extension of the initial LCA ideas is accompanied by an increasing level of standardization. For instance, so called product category rules define frameworks for particular groups of products like dairy in order to ensure comparability of LCA studies in that group. Environmental product declarations are the standardized documents of such assessments. The before mentioned developments are consequences of academic progress and the continuous growth of LCA within business. Nowadays, large car manufacturers, chemical companies, plastics associations or the construction sector use and disclose LCA studies (or excerpts thereof) routinely and therefore require a feasible degree of standardization. At the same time, LCA-related service companies (consulting, software etc.) are established parts of the global business world. LCA is taught at universities and has become a core expertise of academic faculties, institutes, and think tanks in fields such as industrial ecology, sustainability assessment, environmental engineering and so forth. In larger research settings, especially within EU funding schemes, LCA studies are often mandatory and an integral part of accompanying research.

Concluding that LCA is therefore fully matured appears to be exaggerated, though. Within the core methodology issues such as allocation, uncertainty analysis, or comparison and weighting of environmental impact categories leave room for further improvement. Data gaps and data quality can still be troubling, especially when it comes to geographically distinct information. Social LCA and the integration of environmental, social, and economic LCA to form a sustainability LCA are in their early stages. The search for the best system representation in large-scale settings has led to discussions of attributional vs. consequential modeling. The digitalization and big data age raises questions of automated LCA and full integration into enterprise resource planning systems and other IT tools. Paradigms like circular economy and cradle-to-cradle match life cycle thinking well, but the exact interplay leaves room for further research and development.

4 Role of Research and Young Researcher Networks

While further standardization is industry's consequence to the challenges above, increased research activities are an obvious academic response. LCA research necessitates interdisciplinary approaches, which is true for the fundamental and methodological challenges as well as for the challenges in direct application contexts. Accordingly, LCA research does not only take place at a few specialized institutes for life cycle thinking or industrial ecology, but at many rather different organizations and institutes, of which some focus on thermodynamics and process engineering while others are rooted in sociology or forestry. LCA is common ground for some of their research activities and provides opportunities for mutual learning. In many of such cases, LCA knowledge acquisition happens on-the-job or "on-the-PhD", so that peer groups to discuss ideas and challenges are highly important.

One such peer group is the "Ökobilanzwerkstatt" (translates to "LCA workshop", see http://www.oekobilanzwerkstatt.tu-darmstadt.de). This workshop gathers PhD students and further young LCA researches from German-speaking countries and provides room for in-depth discussions of their ongoing research including inputs from senior experts and practitioners in the field. Coordinated by Darmstadt Technical University, 13 annual workshops took place since 2005 at various universities and institutes. These activities have supported interdisciplinary LCA research within German-speaking countries tremendously and initiated the development of further regional and informal LCA researcher networks. Gathering experience in publishing peer-reviewed LCA papers is a further goal of the network. For instance, previous workshop organizers edited a special issue of the journal Sustainability Management Forum (SMF 2016) to publish the participants' short papers after a peer-review process. This current book comprises peer-reviewed contributions from the 2017 Ökobilanzwerkstatt at Technische Universität Braunschweig and demonstrates the interdisciplinarity and diversity of LCA research.

References

EC-JRC—European Commission-Joint Research Centre: ILCD Handbook: General Guide for Life Cycle Assessment—Provisions and Action Steps. EC-JRC (2010)

Hauschild, M., Rosenbaum, R.K., Olsen, S.: Life Cycle Assessment. Theory and Practice. Springer, Cham (2018)

ISO 14040:2006: Environmental management—life cycle assessment—principles and framework. International Organization for Standardization

ISO 14044:2006: Environmental management—life cycle assessment—requirements and guidelines. International Organization for Standardization

SMF—Sustainability Management Forum 1/2016: SMF replaced the former uwf – UmweltWirtschaftsForum

Part II
New Methodological Developments

Using Network Analysis for Use Phase Allocations in LCA Studies of Automation Technology Components

Mercedes Barkmeyer, Felipe Cerdas and Christoph Herrmann

Abstract This article discusses the consideration of the use phase of automation technology components (ATCs) in Life Cycle Assessment (LCA). It aims to contribute to the LCA methodology by exploring a relevance-based allocation method for assessing ATCs and pneumatic components respectively. Although the environmental impact of ATCs in the use phase is rather insignificant when analyzed as a single component, its interaction with other components within an application implies further leverages to the impact caused by the system in the use phase. The social network analysis is an interdisciplinary method that is used in empirical social research and well likely to close the gap between micro level and macro level. This article transfers the social method to automation technology components aiming at estimating the specific contribution of each component to the overall system impact in the use stage.

Keywords LCA · Sustainable components · Allocation use phase
Interaction · Network analysis

1 Introduction

Life Cycle Assessment (LCA) is a system analysis methodology that allows the quantification of potential environmental impacts of product systems (Hauschild et al. 2018). One important challenge when applying LCA for the assessment of complex products is that the environmental impact (e.g. the global warming potential) is only attributed to the actively energy consuming components of the system

M. Barkmeyer (✉)
Festo AG & Co. KG, Ruiter Straße 82, 73734 Esslingen, Germany
e-mail: mercedes.barkmeyer@festo.com; m.barkmeyer@tu-braunschweig.de

M. Barkmeyer · F. Cerdas · C. Herrmann
Chair of Sustainable Manufacturing & Life Cycle Engineering, Institute of Machine Tools and Production Technology (IWF), Technische Universität Braunschweig, Langer Kamp 19b, 38106 Brunswick, Germany

© Springer Nature Switzerland AG 2019
L. Schebek et al. (eds.), *Progress in Life Cycle Assessment*, Sustainable Production, Life Cycle Engineering and Management,
https://doi.org/10.1007/978-3-319-92237-9_2

(e.g. a pneumatic cylinder). While this simplifies the assessment of the influence of the product during the use phase, a more detailed analysis of the contribution of particular parts to the overall performance of the product as a system might contribute enhancing product design and development and might result in a fairer assessment of the individual components.

A complex product architecture can be considered as a network of components that sharing technical interfaces so as to deliver a particular functionality (Sosa et al. 2007). For the case of ATCs, the impact of the use phase of pneumatic components is of high interest because almost 80% of the environmental burdens are caused in this life cycle phase (Festo AG & Co. KG 2015). Components are designed to interact with others. Understanding these interfaces can result in meaningful insights that may contribute to adopt different perspectives for sustainable innovations and business models. The following paper presents a concept how to use the social network analysis in LCA to analyze the contribution of specific component to the overall performance of the system during the use phase. An important advantage of the social network analysis is, that it provides quantitative indicators, that can be integrated as factors in order to allocate the impacts of the use phase to each particular component of the product architecture.

2 Network Analysis

Network analysis is an interdisciplinary method with explanatory power for properties at the macro level and breaks them down to the micro level. Supervising individuals at the micro level is possible—however, this basis is not sufficient to draw reliable conclusions about the society because the interactions of individuals are not considered. At the macro level additional properties, in other words, interaction between individuals exist (emergent properties). A sentence that is often used to describe this fact is: "The whole is more than the sum of its parts" (Jansen 1999). Due to its hybrid position between individuals and society, the social network analysis is generally well suited to close the gap between any micro and macro level (see Fig. 1). For this reason, the method can be applied to analyse emergent properties of applications (in the sense of production systems/composed systems of various components).

3 Concept

As part of research for sustainable components, a framework is already set up by (Barkmeyer et al. 2016), which includes the perspectives of stakeholders (company, customer, society). Integrating this with two general approaches "reduction of environmental damage" and "provide environmental value added", this leads to six categories which can be seen in Fig. 2. Based on this understanding a component with a good LCA result (minimizing damage) that is sold into an industry sector that produces endproducts with controversy environmental and social impacts (dirty business) is not sustainable yet. To achieve the status of a sustainable component, all six categories (Fig. 2) must be considered. Suppliers of components (company) are in a responsible position within industrial automation because their innovations and decisions have transformative potential for production systems (customer) and the produced end-products (society) consequently. As a basis for the category "Target zero damage" (see Fig. 2), the proposed allocation method in this concept helps to describe the environmental damage for a component in the customer's application. The method contains five steps that are described in the following.

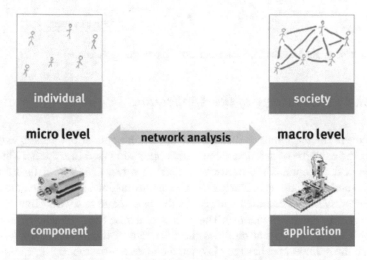

Fig. 1 Transfer of social network analysis to components

Fig. 2 Framework with six categories for defining sustainable components

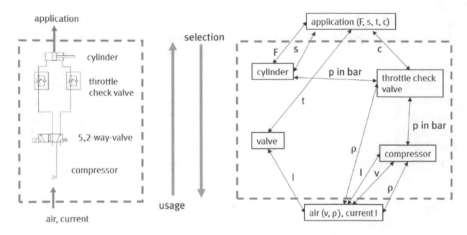

Fig. 3 Exemplary application as use case and modeling of the application

3.1 Step 1—Modeling of the Application

Step 1 covers the modeling of the connections in an application. This is explained
based on the example of a simple pneumatic application (see Fig. 3 left). The pneumatic application contains a compressor (Co), a 5,2-way-valve (V), two throttle check
valves (T) and a cylinder (C). During use phase current and air enter the compressor
where air pressure is generated and passes the pneumatic 5/2-way-valve (contains
five connections and two switches). The air passes a hose (neglected here) and enters
the throttle check valve, that modifies the flow rate of the system. Finally, the air
flows into the cylinder, pushes the piston and causes a movement.

Which connections exist in an application is based on technical-physical understanding, which the user of the method must consider oneself. For this reason, this
step reminds of the life cycle inventory of the LCA method.

When selecting the components, the nodes in the modeling of the application
are directed in the opposite direction. The system sizes are the same. Therefore,
in modeling, arrows are going in both directions that can be named as reciprocal
relations. The lifetime of each component is assumed as equal and therefore neglected
here.

3.2 Step 2—Ecological Relevance of System Sizes
and Component (ERSS, ERC)

In step two the system sizes of the connection modeling are weighted due to their
ecological relevance (ERSS). This is primarily based on engineering understanding.

Table 1 Weighting of ecological relevance of system sizes (ERSS)

System size	Ecological relevance of system size (ERSS)	System size	Ecological relevance of system size (ERSS)
Force (F)	1	Distance (s)	1
Pressure (p)	2	Cycle time (t)	1
Current (I)	2	Flow velocity (c)	2
Density (ρ)	1	Flow volume (v)	2

	out	C	V	T	Co	in
out	0	1	1	1	0	0
C	1	0	0	1	0	0
V	1	0	0	0	0	1
T	1	1	0	0	1	0
Co	0	0	0	1	0	1
in	0	0	1	0	1	0

	out	C	V	T	Co	in
out	0	F,s	t	c	0	0
C	F,s	0	0	p	0	0
V	t	0	0	0	0	I
T	c	p	0	0	r	0
Co	0	0	0	p	0	I,ρ,v
in	0	0	I	0	I,ρ,v	0

	out	C	V	T	Co	in
out	0	2	1	2	0	0
C	2	0	0	2	0	0
V	1	0	0	0	0	2
T	2	2	0	0	2	0
Co	0	0	0	2	0	5
in	0	0	2	0	5	0
Σ	5	4	3	6	7	7

Connection exists: 1
No connection exists: 0

Defining the connections and Summing up of connections

Summing up of ecological relevance per component

Fig. 4 Transferring ecological relevance of system sizes (ERSS) to ecological relevance of component (ERC)

Force, distance, cycle time und air density have small ecological relevance, whereas flow velocity, air pressure, volume flow, current can be assumed with high ecological relevance due to their direct influence on energy demand (Table 1).

For further calculations, an adjacency matrix (see Fig. 4, left) is set up. The adjacency matrix describes whether a connection exists ($a_{ji} = 1$) or not ($a_{ji} = 0$). After that, each system size is multiplied with its ecological relevance of system size (ERSS) and summed up per connection (e.g. ERSS(F)*1 + ERSS(s)*1 = 2). The sum per column is the ecological relevance per component (ERC) (see $\sum$ Fig. 4, right matrix) and can that way be related to the component.

3.3 Step 3—Importance of the Component (IC)

As the third step, the interdisciplinary method social network analysis is used to identify the importance of the component within the system. To determine the quantitative indicators of the social network analysis, various programs can be used (e.g. UCINET, SONIS, R software) (Jansen 1999). For this case study the statistical analy-

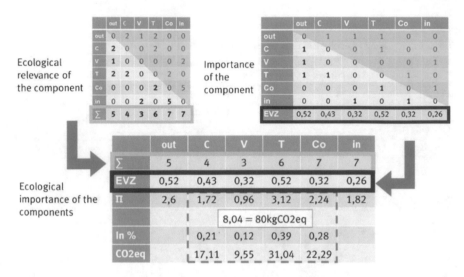

Fig. 5 Calculation of ecological importance of the component (EIC)

sis was performed using R software. The most suitable indicator of the social network analysis, which indicates the importance of each component based on the number of connections is the eigenvector centrality. For the calculation of the eigenvector centrality per component, the adjacency matrix, that was generated after modeling the connections between the components in step 2, serves as input data.

3.4 Step 4—Ecological Importance of the Component (EIC)

For allocating the environmental impact of the application to each component, the ecological importance of the component (EIC) is defined as an allocation factor. This factor can be determined by multiplying the ecological relevance of the component (ERC) with the importance of the component in the application (IC). EIC = ERC * IC. Knowing that the application (cylinder, valve, throttle check valves and compressor) causes 80 kg CO_2eq (as an exemplary assumption) during use phase, the ecological importance can be calculated by applying the rule of three. The 80 kg CO_2eq can be split accordingly to the allocation factor and the EIC respectively. This procedure is illustrated for the exemplary use case in Fig. 5.

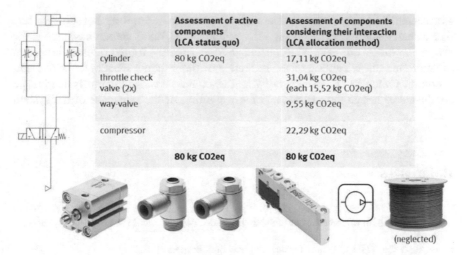

	Assessment of active components (LCA status quo)	Assessment of components considering their interaction (LCA allocation method)
cylinder	80 kg CO2eq	17,11 kg CO2eq
throttle check valve (2x)		31,04 kg CO2eq (each 15,52 kg CO2eq)
way-valve		9,55 kg CO2eq
compressor		22,29 kg CO2eq
	80 kg CO2eq	**80 kg CO2eq**

(neglected)

Fig. 6 Comparison of exemplary results of LCA status quo versus LCA allocation method

4 Results

The results of the employed allocation method are presented in Fig. 6. The differences between the LCA status quo and the allocation method become obvious. The LCA status quo allocates the environmental impact of the air pressure demand only to the active component, the cylinder (80 kg CO_2eq), whereas the throttle check valves and the way-valve have no environmental impact during use phase. Using the here introduced allocation method and considering the interactions of the components, each component is responsible for a certain environmental impact proportional depending on its ecological importance within the system (EIC): each throttle check valve mathematically contributes with 15.52 kg CO_2eq to the environmental impact of the application; the way-valve contributes with 9.55 kg CO_2eq. This implicates, that research and system innovation should focus on the most environmental relevant connection during use phase.

5 Conclusion

This article provides an allocation method to allocate environmental burdens in the use phase of passive components which has not been found in current LCA studies yet. The relevance of the assessment of connections and interactions is clearly highlighted by the results. In order to understand the performance of one component in an application it is necessary to analyze the whole system. For this purpose, the social network analysis was used. Overall, this study strengthens the approach that the network analysis is a suitable methodology to offer new perspectives in life cycle assessment. Since the conceptual allocation method is descriptive, hot spots for further research and potential system innovation can be found. Recommendations for design improvements cannot be concluded. More broadly, further research is needed to determine the weighting of the relevance of the systems sizes in the social and economic dimension of sustainability. In this context a sensitivity analysis must be conducted to hedge the method and the weighting of the relevance of the system sizes.

References

Barkmeyer, M., Kaluza, A., Pastewski, N., Thiede, S., Herrmann, C.: Integration of stakeholder perspectives for development of sustainable automation components. Procedia CIRP 48, 388–393 (2016)

Festo AG & Co. KG: LCA study for material selection, Stuttgart (2015)

Hauschild, M.Z., Rosenbaum, R.K., Olsen, S.I.: Life Cycle Assessment Theory and Practice (2018)

Jansen, D.: Einführung in die Netzwerkanalyse: Grundlagen, Methoden, Anwendungen. Wiesbaden, Springer Fachmedien Wiesbaden GmbH (1999)

Sosa, M.E., Eppinger, S.D., Rowles, C.M.: A Network approach to define modularity of components in complex products. J. Mech. Des. **129**(11), 1118 (2007)

Eco-indicators of Machining Processes

Sandra Eisenträger and Ekkehard Schiefer

Abstract Products have a variety of environmental impacts throughout their life cycle. To conserve resources and to protect the environment these impacts should be reduced to a minimum. The easiest way to reduce the environmental impacts of a product is to do it at an early point of the product development process because at this time there are many opportunities to change something without increased costs. At this time, however, it is often too complex to carry out a complete life cycle assessment (LCA) of the product to identify the ecological weak points. Therefore, EcoDesign-methods are used during product development to assess the expected environmental impacts to minimize them over the whole product life cycle. In the research project EcoScreen manufacturing processes were investigated to generate a reliable database for their life cycle inventory (LCI). These LCI datasets were used to generate simple to use eco-indicators to estimate the environmental impacts without carrying out a complete LCA. For the first time, such eco-indicator values, considering all the relevant environmental effects that are generated during these processes, have been created for the machining of parts.

Keywords LCA · Ökobilanzwerkstatt · EcoDesign · Eco-indicators

The research project "Kurzbilanzierung von Fertigung und Abfallbehandlung beim EcoDesign (EcoScreen)" was financed by the program "Forschung für die Praxis" from the Hessen State Ministry for Higher Education, Research and the Arts. Partners were the Frankfurt University of Applied Sciences, Darmstadt University of Applied Sciences and the company e-hoch-3, Darmstadt.The research project "Kurzbilanzierung von Fertigung und Abfallbehandlung beim EcoDesign (EcoScreen)" was financed by the program "Forschung für die Praxis" from the Hessen State Ministry for Higher Education, Research and the Arts. Partners were the Frankfurt University of Applied Sciences, Darmstadt University of Applied Sciences and the company e-hoch-3, Darmstadt.

S. Eisenträger (✉) · E. Schiefer
Laboratory of Product Development and EcoDesign, Frankfurt University of Applied Sciences,
Nibelungenplatz 1, 60318 Frankfurt, Germany
e-mail: eisentraeger@fb2.fra-uas.de

L. Schebek et al. (eds.), *Progress in Life Cycle Assessment*, Sustainable
Production, Life Cycle Engineering and Management,
https://doi.org/10.1007/978-3-319-92237-9_3

1 Life Cycle Inventory (LCI) of Machining Processes

A CNC machine tool must be supplied with electrical power, cutting fluid, pressured air, cutting tools and needs a manufacturing infrastructure. In addition, the work station must be supplied with light, heat, fresh air and various equipments. During machining waste like oily equipment, metal chips and used cutting fluid, as well as emissions are generated. Most of these flows have upstream and downstream processes, i.e. production, cleaning, conditioning and waste treatment of cutting fluid.

The parametric life cycle inventory (LCI) model of machining processes is based on a LCI method for machining processes described by Schiefer (Schulz and Schiefer 1998, 1999a, b; Schiefer 2001; Abele et al. 2005). This LCI model contains the in- and outputs of the machining process and all necessary sub processes (before, after and during machining), like wastewater treatment, treatment of oily scrap, production of machine tools and cutting tools, cleaning of the part after machining, transportation inside the factory, as well as the complete manufacturing infrastructure. The quantity of the different energy, material, waste and emission flows were defined as functions of geometry and process parameters. Factory layout and process planning were setup in terms of technical feasibility, economical reasonability and work regulation aspects. The modeling principles, the containing in- and outputs and the structure of this parametric LCI model are described in Schiefer et al. (2017). The parametric LCI model was created using the professional software openLCA (Green-Delta GmbH, Berlin). Datasets from the ecoinvent database (ecoinvent Association, Zürich, CH) were used for the background operation, i.e. the generating of electricity.

2 EcoDesign

The product development has an important influence on the environmental effects of a product, due to the basic characteristics/properties of the product and its production processes they are directly and indirectly defined during this phase. If product designers would have a practicable and simple tool to compare variants of a part or different production strategies of the same part, they could more easily choose the ecologically compatible one. This would avoid, that ecological weak points occur later on and could prevent additional costs and engineering time. It is important to have indicators that are simple to handle and interpret. In this study the eco-indicator $(EI \, [\text{points}])$, formed according to the single score indicator of the ReCiPe-method [develop by (Goedkoop et al. 2013)], was chosen to indicate the environmental effects of the processes. The normalization is calculated using the hierarchist perspective with average weightings (Goedkoop et al. 2013).

The following description contains only the ecological investigation of the machining of metal parts on CNC turning centers that allows turning, milling and drilling. Machining centers that are primarily used for milling are still being investigated. They have different energy, pressured air and tool consumption as well as other process parameters and strategies.

During the computer aided design of a product and its workpieces the weight of the metal part $(m_{BT} \, [\text{kg}])$, the weight of removed metal $(m_{SP}[\text{kg}])$ and the surface of the final part $(A_{BT}[\text{dm}^2])$ can easily be derived.

A complete process planning is necessary to define the other parameters, as will be discussed hereinafter.

3 Application of the Parametric LCI Model of Machining Processes to Generate Eco-indicators

The amount of all in- and outputs of the LCI model are variable depending on the geometry and material of the workpiece and the process times. Different materials need different process parameters like cutting speed $(v_c[\text{m/min}])$, feed per revolution $(f[\text{mm}])$, and depth of cut $(a_p[\text{mm}])$, which form the volume removal rate. Most material is removed from a workpiece by roughing with high values for feed per revolution and depth of cut. The last cut, called finishing, removes less material with low feed per revolution to get a good surface quality and dimensional accuracy. As a first approximation it can be assumed that the complete surface generated by machining is finished after roughing.

The environmental effects, which were calculated with the parametric LCI model, depend on the chosen material and the process defining parameters. These parameters are the weight of the metal part $(m_{BT}[\text{kg}])$, the weight of the removed metal $(m_{SP}[\text{kg}])$, the surface of the final part $(A_{BT}[\text{dm}^2])$, the transport volume of the blank part $(V_{RT}[\text{dm}^2])$, the stock removal energy $(E_{th}[\text{MJ}])$, the cutting time $(t_h[\text{min}])$, the running time of the machine including times for tool change, fast movement and movement without cutting $(t_g[\text{min}])$ and the time the machine is occupied including setup and handling times $(t_b[\text{min}])$. Figure 1 shows one of the exemplary workpieces with all mentioned parameters generated by carrying out a complete planning of the machining process.

The eco-indicators (EI) can be calculated by defining these parameters and running the LCI-model in the professional software openLCA. This is done with the impact assessment method ReCiPe and normalization with hierarchic weightings.

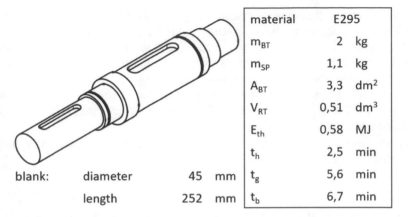

material	E295	
m_{BT}	2	kg
m_{SP}	1,1	kg
A_{BT}	3,3	dm²
V_{RT}	0,51	dm³
E_{th}	0,58	MJ
t_h	2,5	min
t_g	5,6	min
t_b	6,7	min

blank: diameter   45 mm

length   252 mm

Fig. 1 Example workpiece "gear shaft" with all relevant parameters used in the parametric LCI model

Table 1 Conversion factors for machining of an unalloyed steel (E295)

Conversion factors	a	b	c	d	e	f	g	h
E295	0.001	0.0222	0.4450	0.6863	0.0266	0.0024	0.0064	0.0006
Unit	$\frac{\text{points}}{\text{dm}^2}$	$\frac{\text{points}}{\text{MJ}}$	$\frac{\text{points}}{\text{kg}}$	$\frac{\text{points}}{\text{kg}}$	$\frac{\text{points}}{\text{min}}$	$\frac{\text{points}}{\text{min}}$	$\frac{\text{points}}{\text{min}}$	$\frac{\text{points}}{\text{dm}^3}$

4   Investigation of Parameters

Only the material, the weight of the part, the weight of the removed metal and the surface of the final part are known without doing a complete process planning of the machining process. Therefore, the number of parameters had to be reduced to make the eco-indicators applicable during the product development.

By applying the LCI model to different workpieces, a functional relation between the geometry of the workpiece and its environmental effects (eco-indicators) can be identified. This is especially caused by the functional relationship between the geometry of the workpiece and the process times.

The amount of eco-indicators is calculated by Eq. (1). The characters a to h are conversion factors to convert parameters into the eco-indicators (EI). They were defined through parameter studies in the parametric LCI model in openLCA for each investigated material. Table 1 shows the amount of the conversion factors for the unalloyed steel E295 and thereby the environmental impact of the geometry and process parameters for this material.

$$EI = a \cdot A_{BT} + b \cdot E_{th} + c \cdot m_{BT} + d \cdot m_{SP} + e \cdot t_b + f \cdot t_g$$
$$+ g \cdot t_h + h \cdot V_{RT} \quad \text{[points]} \tag{1}$$

The surface of the final part $(A_{BT}[\text{dm}^2])$, the weight of the part $(m_{BT}[\text{kg}])$ and the weight of the removed metal $(m_{SP}[\text{kg}])$ can be easily extracted while using computer aided design and, can therefore, be easily used in the equation. The stock removal energy $(E_{th}[\text{MJ}])$ per kg removed metal is nearly a material specific constant for each investigated material (unalloyed steel, low-alloyed steel, chromium steel 18/8, cast iron, aluminium alloy). For example, for unalloyed steel (E295) the arithmetic averages of the investigated workpieces are 0.47 MJ per kg metal removed by roughing and 0.55 MJ per kg metal removed by finishing.

The total cutting time $(t_h[\text{min}])$ is the sum of the cutting time for roughing $(t_{h_{rou}}[\text{min}])$ and the cutting time for finishing $(t_{h_{fin}}[\text{min}])$. The cutting time per kg removed metal depends mainly on the generated surface and the total removed metal for both roughing and finishing. In each case the minimal value for the cutting time per kg removed metal is reached when a lot of metal is removed to generate a workpiece with minimal surface area. To take these relations into account Eq. (2) can roughly estimate this for each material, either for roughing or for finishing. The equation deduced from investigations of different reference workpieces and their process plans.

$$t_{h_{rou/fin}} = f(x) = \alpha_{rou/fin} \cdot (x)^2 + \beta_{rou/fin} \cdot x + \gamma_{rou/fin} \qquad [\text{min}]$$

$$\alpha, \beta, \gamma := material\ specific\ values\ [min] \tag{2}$$

The variable x represents the ratio of the generated surface to the mass of the removed material. It is also the mass ratio of the material removed by roughing $(m_{SP_{rou}})$ and the total removed material (m_{SP}) and can be calculated by Eq. (3), which only depends on the geometry parameters A_{BT} and m_{SP} and the material specific parameters depth of cut for finishing $(a_{p_{fin}})$ and the density of the material (ρ).

$$x = 1 - \frac{\rho \cdot a_{p_{fin}} \cdot A_{BT}}{m_{SP}} = \frac{m_{SP_{rou}}}{m_{SP}} \qquad [-] \tag{3}$$

The other process times (t_b, t_g) consist of the cutting time and a time slice that is approximately constant for the investigated materials. This is a rough estimation to include auxiliary times like tool changing times, handling times and machine set-up times. Figure 2 shows the value of eco-indicator points per kg removed metal according to the mass of the removed metal and the generated surface using the described relation and the previously described Eq. (3).

These calculated relations could already be used in the product development process. However, the environmental impacts of the manufacturing operation are only one part of the total environmental effects generated by the production of the part. Not yet included in the equation is the production of the not removed material — the weight of the final product. Table 2 containing the eco-indicator values per kg workpiece for three of the investigated materials can be used to take this into account.

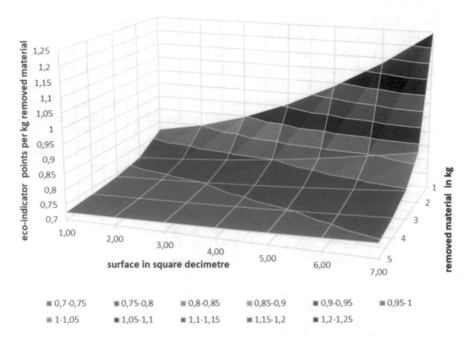

Fig. 2 Relation between the mass of the removed material, the surface, and the resulting eco-indicator points for E295

Table 2 Eco-indicator points for material production and forming per kg of the final product depending on the type of material (generated using ecoinvent data)

Material	$\frac{Points}{kg\ workpiece}$
E295	0.44
1.4301	2.18
42CrMo4	0.63

The final result representing the environmental influence for the production of the part is the summation of the values from Table 2 multiplied with the weight of the part and values taken from the Figs. 3, 4 and 5 depending on the material. These figures show the EI points per kg removed material for the machining process depending on the surface area.

5 Example

In the following, the principle procedure is shown by using three simple work-pieces. They are quite similar in terms of section modulus of torsion W_t $\left(W_{t,1} = 5300 \text{ mm}^3; \quad W_{t,2} = W_{t,3} = 5200 \text{ mm}^3\right)$ and consist of unalloyed steel

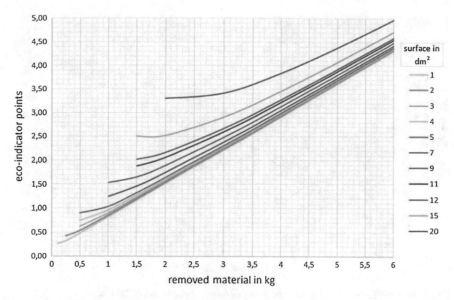

Fig. 3 Eco-indicator points for the machining of unalloyed steel (E295)

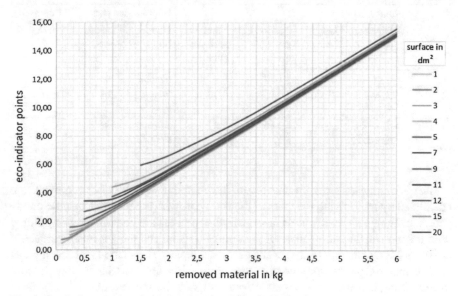

Fig. 4 Eco-indicator points for the machining of stainless steel (1.4301)

(E295). The first one is a shaft, the second one and the third one are hollow shafts. All three have constant diameters over the entire length. Table 3 shows the relevant properties.

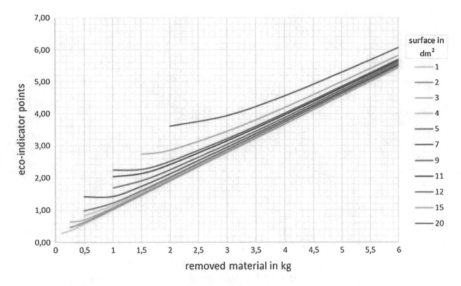

Fig. 5 Eco-indicator points for the machining of low-alloyed steel (42CrMo4)

Table 3 Overview of the three example workpieces (E295)

No.	Final part				Unprocessed part		Removed
	Diameter (mm)	Length (mm)	Mass (kg)	Machined surface (dm^2)	Diameter (mm)	Length (mm)	Mass (kg)
1	$d = 30$	200	1.12	2.03	$d = 32$	205	0.16
2	$d_a = 40$ $d_i = 35$	200	0.46	4.77	$d_a = 42$ $d_i = 32$	205	0.47
3	$d_a = 40$ $d_i = 35$	200	0.46	2.57	$d_a = 42$ $d_i = 35$	205	0.22

The eco-indicators for production of the material, which is not removed, are calculated by multiplying the mass of the final part with the material specific points shown in Table 2. The effects of the machining process are determined using Fig. 4 with the information of the mass of the removed metal and the generated surface.

Table 4 is a summary of the results of the examples. The shaft (1) has the highest score in material production caused by the weight of the final part, but the lowest score in machining caused by the less removed material and the less generated surface. The hollow shaft with the machined inner surface (2) has a lower score in material production, however, it has a higher score in machining caused by the high values for machining of both sides. The values for machining the hollow shaft without machining the inner surface (3) are more similar to the shaft (1) values, but the production of the not lost material has less eco-indicator points. In addition, it is the part with the lowest environmental impact.

Table 4 Summary of the results

	Eco-indicator points		
	Shaft (1)	Hollow shaft (2)	Hollow shaft (3)
Material production (excl. removed material)	0.5	0.2	0.2
Machining (incl. production of removed material)	0.45	0.9	0.5
EI total	0.95	1.1	0.7

6 Prospects

The generated method to estimate the environmental impact of machined parts is simple to handle and is based on detailed investigations of the machining processes and economical reasonable process planning. The parametric structure of the LCI model allows investigations of other materials or of the influence of the number of same pieces.

In the future, other manufacturing processes like welding, forging, water jet or laser cutting, assembling and disassembling, different types of moulding are going to be investigated to generate a database with many different options for production of a product or component.

If the database will contain more processes and different materials, it could allow the comparison of a lot of different variations of geometries and manufacturing processes or production strategies. By applying the method to the product development process, the environmental impact could be reduced efficiently without a distinct increasing of development costs and time.

References

Abele, E., Anderl, R., Birkhofer, H.: Enviromentally-Friendly Product Development—Methods and Tools. Springer-Verlag, London (2005)

Goedkoop, M., Heijungs, R., Huijbregts, M., de Schryver, A., Struijs, J., van Zelm, R.: ReCiPe 2008—A Life Cycle Impact Assessment Method Which Comprises Harmonised Category Indicators at the Midpoint and the Endpoint Level. Ministerie van Volkshuisvesting, Ruimtelijke Ordening en Milieubeheer (NL) (2013)

Schiefer, E.: Ökologische Bilanzierung von Bauteilen für die Entwicklung umweltgerechter Produkte am Beispiel spanender Fertigungsverfahren, Aachen: Shaker (Darmstädter Forschungsberichte für Konstruktion und Fertigung), zugl. Dissertation TU Darmstadt 2000 (2001)

Schiefer, E., Eisenträger, S., Steinberg, I., Kutschmann, J.: EcoScreen – Ökobilanzierung von Fertigungsverfahren. ZWF – Zeitschrift für wirtschaftlichen Fabrikbetrieb. **112**(11), 727–730 (2017)

Schulz, H., Schiefer, E.: Prozessführung und Energiebedarf bei spanenden Fertigungsverfahren. ZWF – Zeitschrift für wirtschaftlichen Fabrikbetrieb. **93**(6), 266–271 (1998)

Schulz, H., Schiefer, E.: Methode zur ökologischen Bilanzierung der Zerspanung. wt – Werkstattstechnik. **89**(5), 239–243 (1999a)

Schulz, H., Schiefer, E.: Methodology for the Life Cycle Inventory of Machined Parts. In: Production Engineering, vol. VI/2, pp. 121–124 (1999b)

Enhancing the Water Footprint Method to a Region Specific Management Tool

Natalia Finogenova, Markus Berger and Matthias Finkbeiner

Abstract Water Footprint (WF) is broadly applied as a method to quantify impacts associated with the water use throughout the value chain of products, nevertheless the need for a more temporally and spatially explicit evaluation has recently been highlighted. In this paper a region specific WF inventory and midpoint impact assessment for the cotton-textile value chain in Pakistan is introduced. The locally relevant parameters are identified and included into the water consumption inventory and water availability database. The results are applied to a numerical model for the cotton cultivation. For the water consumption, the introduced region specific parameters are position on the irrigation channel, water source (distinguishing between surface and groundwater), use of the storage reservoirs and water trade between farmers for the inventory. Parameters groundwater level, and salinity and distinguishing between surface and groundwater are included into the water availability database. The calculated WF demonstrates that the separate assessment of the surface and groundwater in both inventory and impact assessment is essential on a regional level. Evaluating local conditions play the vital role for a robust quantification of the WF. Further development of the region specific impact assessment is needed in particular for the endpoint impact assessment for the areas of protection human health, ecosystems and freshwater resources.

Keywords Water footprint · Region specific impact assessment
Cotton-textile value chain

1 Introduction

Freshwater is essential for human well-being and ecosystems. However, more than 40% of the world's population is living nowadays under acute water stress (OECD

N. Finogenova (✉) · M. Berger · M. Finkbeiner
Chair of Sustainable Engineering, Technische Universität Berlin, Straße des 17. Juni 135, 10623 Berlin, Germany
e-mail: Natalia.finogenova@tu-berlin.de

© Springer Nature Switzerland AG 2019
L. Schebek et al. (eds.), *Progress in Life Cycle Assessment*, Sustainable
Production, Life Cycle Engineering and Management,
https://doi.org/10.1007/978-3-319-92237-9_4

27

2012). According to the Organisation for Economic Co-operation and Development (OECD 2012) the global water demand will increase by 55% between 2000 and 2050, driven by the population growth and industrial development, in particular in emerging and developing countries. Growing European demand for the water intensive products, e.g. crops and ores, is aggravating pressure on the water resources in exporting countries (Ercin et al. 2016). Increasing water scarcity and water quality alteration cause severe health effects, whereas low and middle income level population groups are affected in particular (WHO 2003, 2009). Although water availability is a global issue, distribution of the water resources is characterized by significant regional differences. As demonstrated by Loubet et al. (2013), water availability and effects of water consumption can significantly vary even between different sub-watersheds within one river basin. For some regions, temporal variation is the most crucial factor for water availability, e.g. monsoon and dry seasons in Asia. Thus, to evaluate water scarcity impacts adequately, a regional or even local perspective has to be applied (Berger and Finkbeiner 2010, 2013; Quinteiro et al. 2017; Wichelns 2017).

Water Footprint (WF) has been widely applied to evaluate environmental impacts associated with the water consumption and degradation (pollution) since decades (Berger and Finkbeiner 2010; Aivazidou et al. 2016). The method allows to evaluate potential water use related impacts under the life cycle perspective (DIN 2016). The impacts are quantified by multiplying the water consumption and degradation with the characterization factors (CFs) and are expressed in H_2O-eq..[1] The CFs consider the regional water availability, whereas a number of WF models exist providing the CFs on a country or watershed level (Pfister et al. 2009; Berger and Finkbeiner 2010; Kounina et al. 2013; Berger et al. 2014; Pfister and Bayer 2014; Boulay et al. 2017).

So far, the water footprint studies have been conducted for a broad range of goods, including agricultural products, e.g. tomatoes (Chapagain and Orr 2009) and cotton (Hoekstra et al. 2006; Pfister et al. 2009), ores (Northey et al. 2014; Buxmann et al. 2016) and complex industrial products like a car (Berger et al. 2012).

Despite the broad application of WF, some authors claim that the current WF models are insufficient to address water related problems and identify hotspots on the local level (Hess et al. 2015; Wichelns 2017). Several methodological challenges of the water footprint approach have been identified and discussed recently (Berger and Finkbeiner 2010; Quinteiro et al. 2017), in particular the need for the temporally and spatially explicit water consumption inventories and characterization factors.

In order to address some of these challenges, a region specific water footprint is introduced. First, a region specific WF database is established. For this purpose, locally relevant aspects are investigated and included into the database alongside other parameters commonly used for the WF studies. In the next step the WF impact assessment is adapted to a region specific perspective. Within this paper one aspect relevant for the regional impact assessment is investigated—a separate evaluation of the surface and groundwater. The results are demonstrated by means of a numeric model for the cotton production in Pakistan.

---

[1]Here only the water scarcity footprint is described.

The method development is conducted within the project InoCottonGROW. InoCottonGROW is a collaborative project of 26 partners from Germany, Pakistan and Turkey sponsored by BMBF. The project aims at contributing to a sustainable water use along the cotton-textile value chain in Pakistan by enhancing the Water Footprint method to a region specific management tool.

2 Method

2.1 Development of the Region Specific WF Databases

The region specific WF database is split into the water consumption inventory and the water availability database. First, the common parameters used for calculating the WF on a country or watershed level are considered based on the existing WF models. The latter are, for example, evapotranspiration, water withdrawal and discharge, fertilizers application and cotton yield for the water consumption inventory. The parameters included in the water availability database are, among others, surface runoff, groundwater recharge and human water consumption.

In the next step, locally relevant parameters are identified and included into the databases. For the water consumption inventory, they are derived by means of the on-site farm visits in the province Punjab in Pakistan. The aspect identified as mostly relevant is the position of the farm on an irrigation channel. The irrigation systems leads to an unequal distribution of water, so that the farmers on the head of the irrigation channel get more water than farmers on the tails. Further locally relevant aspects are water trade between farmers and usage of the water storage reservoirs; compensation of the surface water scarcity by using the groundwater; loss of yield due to saline groundwater. For these aspects corresponding parameters are included into the database. The results are presented in Fig. 1. For the water availability database, the relevant aspects are determined based on the existing literature concerning water resources in Pakistan. It was identified that the groundwater level and salinity are the limiting factors for the water availability. Furthermore, the water demand is strongly water source specific depending on the user: while agriculture withdraws the surface water, industrial and domestic sectors use only the groundwater. The corresponding parameters included into the water availability database are presented in Fig. 2.

2.2 Region Specific Impact Assessment Model

To evaluate the influence of the region specific parameters introduced in the water consumption inventory and water availability database, the impact assessment is carried out. For this purpose, a simplified numeric model for the product system cotton cultivation is established. The product system belongs to the water user *agriculture*

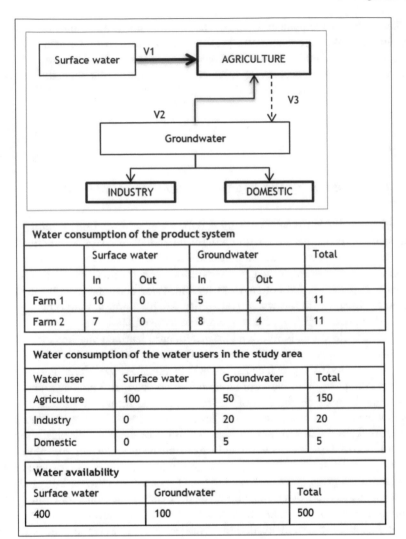

Fig. 1 The numeric model for the calculation of the region specific water footprint for the cotton cultivation including: Water consumption of the product system (cotton cultivation); water consumption of the water users in the study area (agriculture, industry, domestic); water availability in the study area

and withdraws surface (V1) and groundwater (V2) for irrigation, whereas part of the applied water percolates and recharges the groundwater aquifer (V3) (Fig. 2). To evaluate the influence of the parameter *position on the irrigation channel*, the water consumption inventory is modelled for the position on the head of the channel (farm 1) and tail (farm 2). Thus, the first farm gets more surface water and the second farm has to apply more groundwater (see Sect. 2.1). The percolation rate of the water

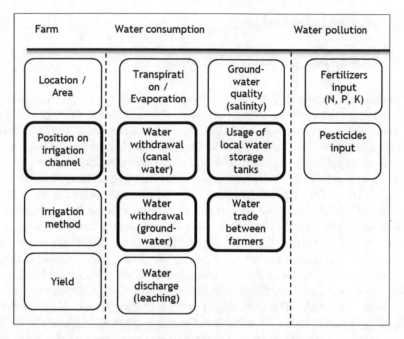

Fig. 2 The region specific water consumption database. The bold frames indicate the introduced region specific parameters

applied to the field is assumed as 30%. The model also includes the users *industry* and *domestic*, which are withdrawing only the groundwater as described in Sect. 2.1.

The water footprint is calculated for each water source (surface or groundwater) by multiplying the water consumption from this water source (WC_i) with the characterization factor (CF_i) for the surface or groundwater, repectively (see Eq. 1). The CF is calculated as the consumption-to-availability ratio. The water consumption of the agricultural, domestic and industrial users is considered.

$$WF_i = WC_i * CF_i = WC_i * \frac{\left(WC_i^{agriculture} + WC_i^{industry} + WC_i^{domestic}\right)}{Water\ availability_i} \quad (1)$$

The total water footprint is then calculated at the sum of the water footprint for the surface and groundwater use. The results are compared to the average WF calculated without distinguishing between different water sources.

The results are presented in the next chapter.

3 Results and Discussion

3.1 The Region Specific WF Database

The parameters included in the water consumption inventory are presented in Fig. 2. The inventory is divided into three data blocks: *farm attributes*, *water consumption* and *water pollution* aspects. In the following, the introduced region specific parameters are described. The farm position on the irrigation channel is included as a new parameter into the block *farm attributes*, since it significantly influences water availability on a farm level as described in Sect. 2.1. Water consumption is divided into the surface (from the irrigation channel) and groundwater, which allows considering different availability of the surface and groundwater in the impact assessment. Furthermore, local measures such as using the water storage tanks and water trade between farmers to balance the water availability are included into the inventory.

The parameters included in the water availability database are presented in Fig. 3. The database is divided into three blocks: *area attributes*, *water availability* and *water consumption* aspects. The introduced region specific parameters are groundwater level and salinity. Both parameters are the limiting factors for groundwater availability in particular for the agricultural sector, on the one hand, due to the high energy costs for water pumping, on the other hand, due to the damage of the saline water when using for the irrigation purposes. The *water consumption* is divided into the surface and groundwater to consider the different water sources used by the agricultural, industrial and domestic sectors.

3.2 The Region Specific Impact Assessment

The results of the numeric model are presented in Fig. 4 and demonstrate how distinguishing between the surface and groundwater influences the WF. The CF for the groundwater availability (0,75) is three times higher than for the surface water (0,25), thus groundwater is three times more scarce than the surface water in the region. This result is caused by the high demand for the groundwater compared to its resources. The average CF (0,35) is significantly lower than the CF for the groundwater and slightly higher than the CF for the suface water. This can lead to over- or under estimation of the water scarcity, in particular when a user, e.g. industrial sector, is consuming water from only one source.

The average WF is equal for both farms (3,9 H_2O-eq.), while, when evaluating surface and groundwater separately, it sums up to 3,3 H_2O-eq. for the first farm and 4,8 H_2O-eq. for the second (Fig. 4). The reason is, that the farmer on the head of the irrigation channel (farm 1) applies mainly surface water, while the farmer on the tail (farm 2) applies more groundwater due to the lack of the irrigation water. Since water scarcity of the groundwater is higher than of the surface water, the WF of the farm 2 is higher as well.

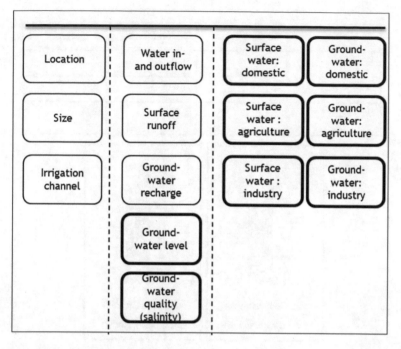

Fig. 3 The region specific water availability database. The bold frames indicate the introduced region specific parameters

4 Conclusion and Outlook

This paper introduces a water consumption inventory and a water availability database for calculating the region specific water footprint on example of cotton cultivation in Pakistan. Five regional parameters are identified and included into the water consumption database: position on the irrigation channel, distinguishing between the surface and groundwater consumption, usage of the water storage tanks and water trade between farmers. For the water availability database the new considered parameters are groundwater level and quality as well as separate evaluation of the surface and groundwater availability. The method is applied to a numerical model. The results demonstrate that water scarcity and thus water footprint can significantly vary for surface and groundwater, which might lead to an under- or over estimation of the water footprint.

The data availability might be a limitation factor for calculating a region specific WF. Further steps include development of the region specific cause-effect chains for the endpoint impact assessment for human health, freshwater resources and ecosystems. Furthermore, the linkages between the WF results and the Sustainable Development Goals indicators might be established to provide monitoring and decision

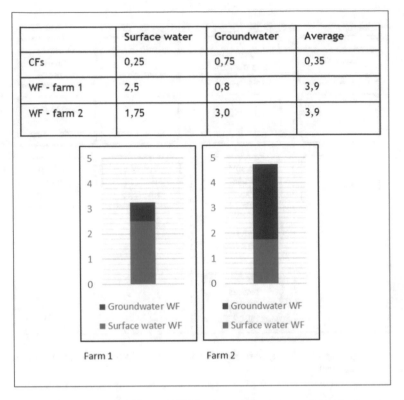

	Surface water	Groundwater	Average
CFs	0,25	0,75	0,35
WF - farm 1	2,5	0,8	3,9
WF - farm 2	1,75	3,0	3,9

Fig. 4 CFs for the water availability and WF for the surface water use, groundwater use and the average WF (not distinguishing between surface and groundwater) for the farm on the head (farm 1) and on the tail (farm 2) of the irrigation channel

support for political decision makers. A robust approach to include the water quality parameters into the WF method needs to be established.

References

Aivazidou, E., et al.: The emerging role of water footprint in supply chain management: a critical literature synthesis and a hierarchical decision-making framework. J. Clean. Prod. **137**, 1018–1037 (Elsevier Ltd) (2016). https://doi.org/10.1016/j.jclepro.2016.07.210

Berger, M., Finkbeiner, M.: Water footprinting: how to address water use in life cycle assessment? Sustainability **2**(4), 919–944 (2010). https://doi.org/10.3390/su2040919

Berger, M., Finkbeiner, M.: Methodological challenges in volumetric and impact-oriented water footprints. J. Ind. Ecol. **17**(1), 79–89 (2013). https://doi.org/10.1111/j.1530-9290.2012.00495.x

Berger, M., et al.: Water footprint of European cars: potential impacts of water consumption along automobile life cycles. Environ. Sci. Technol. **46**(7), 4091–4099 (2012). https://doi.org/10.1021/es2040043

Berger, M., et al.: Water accounting and vulnerability evaluation (WAVE): considering atmospheric evaporation recycling and the risk of freshwater depletion in water footprinting. Environ. Sci. Technol. 48(8), 4521–4528 (2014). https://doi.org/10.1021/es404994t

Boulay, A.-M., et al.: The WULCA concensus characterization model for waer scarcity footprints: assessing impacts of water consumption based on available water remaining (AWARE). Int. J. Life Cycle Assess. (2017). https://doi.org/10.1007/s11367-017-1333-8

Buxmann, K., Koehler, A., Thylmann, D.: Water scarcity footprint of primary aluminium. Int. J. Life Cycle Assess. 21(11), 1605–1615 (2016). https://doi.org/10.1007/s11367-015-0997-1

Chapagain, A.K., Orr, S.: An improved water footprint methodology linking global consumption to local water resources: a case of Spanish tomatoes. J. Environ. Manage. 90(2), 1219–1228 (Elsevier Ltd) (2009). https://doi.org/10.1016/j.jenvman.2008.06.006

DIN: Umweltmanagement – Wasser-Fußabdruck – Grundsätze, Anforderungen und Leitlinien (ISO 14046:2014); Deutsche und Englische Fassung EN ISO 14046:2016 (2016)

Ercin, A.E., Chico, D., Chapagain, A.K.: Dependencies of Europe's economy on other parts of the world in terms of water resources, Horizon 2020—IMPREX project, Technical Report D12.1, Water Footprint Network (2016)

Hess, T.M., Lennard, A.T., Daccache, A.: Comparing local and global water scarcity information in determining the water scarcity footprint of potato cultivation in Great Britain. J. Clean. Prod. 87(1), 666–674 (Elsevier Ltd) (2015). https://doi.org/10.1016/j.jclepro.2014.10.075

Hoekstra, A.Y., et al.: The water footprint of cotton consumption: an assessment of the impact of worldwide consumption of cotton products on the water resources in the cotton producing countries. Ecol. Econ. 60, 186–203 (2006). https://doi.org/10.1016/j.eco%20lecon.2005.11.027

Kounina, A., et al.: Review of methods addressing freshwater use in life cycle inventory and impact assessment. Int. J. Life Cycle Assess. 18(3), 707–721 (2013). https://doi.org/10.1007/s11367-012-0519-3

Loubet, P., et al.: Assessing water deprivation at the sub-watershed scale in LCA including downstream cascade effects. 23th SETAC Eur. Annu. Meet. (2), 2 p. (2013). https://doi.org/10.1111/j.1530-9290.2012.00495.x

Northey, S.A., et al.: Evaluating the application of water footprint methods to primary metal production systems. Miner. Eng. 69, 65–80 (Elsevier Ltd) (2014). https://doi.org/10.1016/j.mineng.2014.07.006

OECD: Water. OECD Environ. outlook to 2050 Consequences Ina. 208 p. (March) (2012). https://doi.org/10.1787/9789264122246

Pfister, S., Bayer, P.: Monthly water stress: spatially and temporally explicit consumptive water footprint of global crop production. J. Clean. Prod. 73, 52–62 (2014). https://doi.org/10.1016/j.jclepro.2013.11.031

Pfister, S., Koehler, A., Hellweg, S.: Assessing the environmental impact of freshwater consumption in life cycle assessment. Environ. Sci. Technol. 43(11), 4098–4104 (2009). https://doi.org/10.1021/es802423e

Quinteiro, P., et al.: Identification of methodological challenges remaining in the assessment of a water scarcity footprint: a review. Int. J. Life Cycle Assess. 1–17 (2017). https://doi.org/10.1007/s11367-017-1304-0

WHO: Domestic water quantity, service level and health. World Health Organ. 39 p. (2003). https://doi.org/10.1128/jb.187.23.8156

WHO: Global health risks: mortality and burden of disease attributable to selected major risks. Bull. World Health Organ. 87, 646–646 (2009). https://doi.org/10.2471/blt.09.070565

Wichelns, D.: Volumetric water footprints, applied in a global context, do not provide insight regarding water scarcity or water quality degradation. Ecol. Indic. 74, 420–426 (2017). https://doi.org/10.1016/j.ecolind.2016.12.008

Product System Modularization in LCA Towards a Graph Theory Based Optimization for Product Design Alternatives

Chris Gabrisch, Felipe Cerdas and Christoph Herrmann

Abstract In light of current environmental challenges, industrial companies are increasingly required to reduce their individual environmental impact. As these companies face economic constraints, the reduction of the specific impacts needs to be achieved in the most cost-efficient manner. This is leading to trade-offs between the potential environmental improvements driven by particular measures and the costs of these measures. Due to the inherent complexity of product systems many different measures to alter the products properties exist, leading to a high number of possible combination alternatives in the foreground system and consequently to many different product system set-ups and LCA results. Modular LCAs are an approach to calculate these results by performing separated LCAs for all individual life cycle modules, which afterwards are reconnected again to form the LCA results for all possible module combinations. However, when the LCA result of one of these modules is influenced by interactions with other modules, the consideration of these influences leads to a fast rise in the data demand for a modular LCA. Modelling such an optimization problem via graph theory can be a possible way to address interdependencies between modules while still being able to provide the necessary data demand through a systematic graph design.

Keywords LCA · Optimization in LCA · Graph theory · Data demand

1 Introduction

To reduce the human interventions with the ecosphere, political agreements regarding the decrease of environmental pollution or global warming have been passed. The Agreement of Paris by the United Nations is a demonstration of the worldwide

C. Gabrisch (✉) · F. Cerdas · C. Herrmann
Chair of Sustainable Manufacturing and Life Cycle Engineering, Institute of Machine
Tools and Production Technology (IWF), Langer Kamp 19 B, 38106 Brunswick, Germany
e-mail: c.gabrisch@tu-braunschweig.de

C. Gabrisch
Volkswagen AG, Berliner Ring 2, 38440 Wolfsburg, Germany

© Springer Nature Switzerland AG 2019
L. Schebek et al. (eds.), *Progress in Life Cycle Assessment*, Sustainable
Production, Life Cycle Engineering and Management,
https://doi.org/10.1007/978-3-319-92237-9_5

understanding for the need to limit the rise of the global mean temperature to well below 2 °C compared to the pre-industrial level (United Nations 2015). For commercial organisations with business activities or products related to high greenhouse gas emissions, it therefore becomes necessary to reduce the environmental impact of their products or services. As the companies are competing in an economic-focused market, it becomes more and more reasonable to decrease the specific impacts in these areas of operation, where it is cheapest (Poppe 2001).

Companies that produce goods with a complex life cycle (e.g. long supply chain, diverse mix of materials and/or energy-intense use phase) therefore need to assess a cost minimal configuration of their product over the whole life cycle which reduces their environmental impact to a given limit. This typical life cycle consists of a production phase, a use phase and an end of life phase of the product (Broch 2017).

To optimize this life cycle for a given environmental target, e.g. regarding the emissions of greenhouse gases, different approaches can be possible. One approach could be a single emission reduction measure in one of the three phases or a combination of multiple measures throughout the whole life cycle.

To identify this ideal configuration of the products life cycle, an optimization algorithm is necessary, which identifies this ideal set of measures for a given environmental target. Such an algorithm needs information about the individual potential of greenhouse gas emission reduction and costs of each measure. These input data can e.g. be provided via life cycle costing and an environmental impact analysis based on the ISO Norm 14040.

2 Modularization of Product Systems in LCA

With a growing complexity of the analyzed product, the calculation of an LCA becomes a bigger effort. If the product should be improved regarding the environmental properties, many different alternatives can occur across every phase of the products life cycle. A high diversity of possible measures to reduce the greenhouse gas emissions leads to a high demand of LCA-results, as each possible life cycle, which is the result of a unique combination of measures, needs to be assessed separately (Herrmann et al. 2013).

If a given products life cycle can be altered with exemplary 75 different measures aiming for environmental improvement, which can be combined in any possible way, the LCA demand reaches with more than $3.78 * 10^{22}$ possible combinations extremely high numbers. The established method of "modular LCA" can help to reduce the calculation effort back to an individual LCA for each of the initial 75 measures. A modular LCA of a product system is based on the concept of dividing the life cycle into clearly separable modules, which are then assessed individually (Jungbluth 2000). The LCAs of the individual modules added up result in the LCA of the whole product system. With a high number of alternative modules (which can be combined in many different ways), the effort to analyze the environmental impact

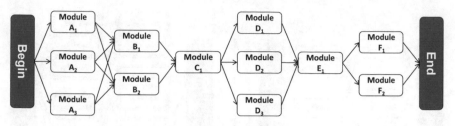

Fig. 1 Modular life cycle model

of only the modules is far lower than it would be to evaluate the whole life cycle each time (Cerdas et al. 2018; Steubing et al. 2016).

Figure 1 shows an exemplary product system with six different life cycle stages (A to F) and one to three different modules per stage.

In total, 36 combinations or alternative life cycles are possible for this example and thus possibly 36 different LCA results for the final product. With modular LCAs, this effort can be reduced to twelve LCAs (one for each module), which can then be combined in all possible ways afterwards. Modular LCAs are therefore a smart approach to reduce the amount of necessary LCAs to compare many different life cycle alternatives.

3 Interdependencies Between Measures

A modular LCA approach might lead to wrong results, when the individual LCA of each module is assumed to be static, but actually depends on the modules, it is combined with. When measures are sensitive to the combination with other measures, the initial concept of modular LCAs cannot be applied, as a measure does not have a fixed or static individual LCA result anymore but multiple values, depending on the combination context with other measures. The ongoing calculation with fixed values for interchangeable measures would then lead to incorrect results. This phenomenon can be described as "interdependencies" between measures, respectively modules. (Herrmann et al. 2013)

Referring back to the example of the 75 measures, the calculation of only 75 LCAs and an utilisation of these results in a modular LCA approach cannot be applied, if possible interdependencies shall be considered. If a measure interacts with other measures, than each measure can possibly have an individual value for each possible combination of measures in which it takes part.

On the left side of Fig. 2, an exemplary life cycle with two stages and three modules for stage A (A_1 to A_3) and two modules for stage B (B_1 and B_2) is displayed.

In total, six alternative life cycles are possible (A_1B_1, A_1B_2, A_2B_1, A_2B_2, A_3B_1, A_3B_2), which can be assessed with only five LCAs (one for each module) via modular LCA. On the right side of Fig. 2 is the same exemplary life cycle depicted.

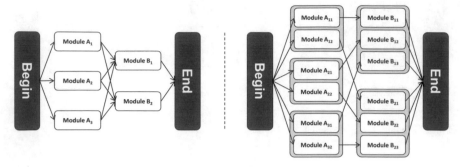

Fig. 2 Interdependencies between LCA modules

The difference is that possible interdepencies are considered. For example, module A_1 can be combined with either module B_1 or module B_2. If the LCA result of module A_1 depends on the exact combination with a certain module of stage B, than module A_1 can have two different LCA results, one for the combination with module B_1 and one for the combination with module B_2. The same logic applies to the other modules. The total number of possible modules which need to be analyzed with an LCA to apply a modular LCA therefore rises from five to twelve for the same six possible life cycles. Hence, the number of necessary modules surpasses the number of possible life cycles. For larger networks, the difference between the amount of alternative life cycles and the necessary data demand for individual modules growths quickly.

This example shows that the consideration of interdependencies between individual modules within a modular LCA leads to an enormous data demand regarding individual LCAs for each module. This leads to the need for a strategy to reduce the data demand for individual modules below the number of possible life cycles without losing the level of detail, which the consideration of the interdependencies provides for the final results.

4 Data Demand Reduction Strategies

The depiction of the modular LCA and the complexity of interdependencies in the previous sections have been explained via graph theory, as this form of problem representation is very suitable to exemplify the logical and combinatorical relations of life cycle stages. Since the interdependencies between these modules arise due to certain combinations within the chain of life cycle stages, changes in the design of the graph are a helpful approach for the reduction of the data demand. The detailed modifications to these graphs are explained in the following chapter.

The number of existing individual modules considering the possible interdependencies can be calculated with Eq. (1):

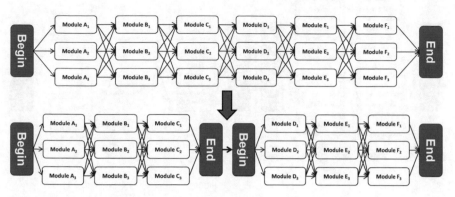

Fig. 3 Forming of subgraphs

$$N = y * \left(n_1 * n_2 * \ldots * n_y \right) \tag{1}$$

N=Number of individual Modules
y=Number of life cycle stages
n_y =Number of alternatives for each life cycle stage

To reduce the result of this equation, either the number of alternatives per stage or the number of stages needs to be reduced. One effective way of reducing the data demand for a network of a whole life cycle therefore is the principle of forming separated subgraphs, which are afterwards connected again. A subgraph in this context is a graph G', where the vertices (V') and edges (E') are all part of the vertices (V) and edges (E) of the main graph G (Domschke et al. 2015). If it is possible to divide a given graph G into two subgraphs G_1' and G_2', the number of possible individual modules reduces quickly, as fewer possible connections appear that lead to the existence of individual modules in G_1' and G_2'.

The top graph in Fig. 3 would have a total amount of 4374 individual modules [see Eq. (2)], while the two separate subgraphs in the bottom have a data demand for 81 individual modules each [see Eq. (3)].

$$4374 = 6 * (3 * 3 * 3 * 3 * 3) \tag{2}$$

$$81 = 3 * (3 * 3 * 3) \tag{3}$$

If every combination of the bottom two graphs are afterwards combined with each other, than both (the top and the bottom) networks offer $3^6 = 729$ possible combinations or alternative life cycles. While the top network needs 4374 individual modules to provide correct results for the LCAs with interdependencies, it becomes also feasible to calculate these results with only 162 individual modules from the bottom network, if it is possible to split the network in half and calculate separated results. This separation of subgraphs leads to a reduction of -96.3% of necessary

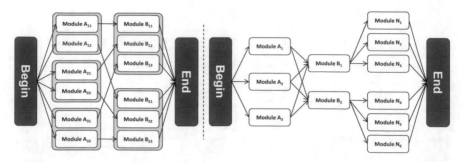

Fig. 4 Excluded stage of interdependencies

individual modules. While the data demand in the top network exceeds the number of possible combinations, the number of individual modules in the bottom network is below that value.

This method is helpful to reduce the necessary LCA data input for a complex life cycle where many, but not all, measures interact. But this approach only helps to reduce the number of individual modules, when multiple subgraphs are combined. Depending on the size of a subgraph, the number of individual modules can also become very large within a single subgraph, which makes it necessary to reduce the amount of data demand not only for the whole life cycle graph, but also within a subgraph itself. The need for the high number of individual modules is based on the possibility of interdependencies between combined measures regarding their specific potential of CO_2-reduction. The source of these emissions, e.g. in the context of a vehicle can be split into the emissions of the production of a measure and the influence of these measures on the emissions during the use phase of the product, if these measures interact.

An additional way of reducing the amount of needed individual modules therefore is to summarize and exclude all the information of interdependencies into one additional stage at the end of the graph. In Fig. 4 on the right side, the graph grows by the stage N, the modules N_1 to N_6 hereby represent the six different interdependency-possibilities of the network. The basic modules A_1 to A_3 and B_1 to B_2 now do not have to be assessed for each interdependency individually, as these basic modules now only store the information of their specific production phase, where the LCA result does not change with further connection of additional measures in this example, while their influence on the use phase are all combined in the newly added stage N. This way, the whole network can provide the result for all combinations including the interdependencies with eleven modules (right side of Fig. 4) instead of twelve modules (left side of Fig. 4). This strategy applied to the two subgraphs in Fig. 3 leads to a number of individual modules of only 36 per subgraph instead of 81, which results in a number of individual modules of 72 for the whole network with a total of 729 possible combinations.

5 Example

An example for the need of environmental optimization within a complex product can be e.g. found in the automotive industry. With the possibility of adapting different materials, light weight design, powertrains and recycling concepts, many different measures can be applied to a cars life cycle. Interferences between life cycle modules hereby can e.g. be different fuel reduction values for a mass reduction of e.g. 100 kg (possibly by a light weight designed carbody) for different powertrains (Koffler and Rohde-Brandenburger 2010). Transferred to the life cycle graph, at some point the stage "carbody" needs to be combined with the stage "powertrain". Due to the different fuel reduction values, the emission reduction potential of the same carbody varies for every powertrain it is combined with. These different emission reduction potentials arise due to the different specific emissions of each powertrain for the provision of 1 kWh of energy. While the weight reduction of 100 kg influences the energy demand in kWh for both powertrains in the same amount, the different powertrains save different amounts of CO_2-emissions due to the reduced energy demand in kWh (Rohde Brandenburger 2013). An additional aspect is the recuperation of energy of electric vehicles, which allows heavier vehicles to retrieve more energy during brake-processes (Vetter 2017). This is why the same lightweight designed carbody has a specific CO_2-influence for every powertrain that it is connected with.

A possible use case for subgraphs, e.g. in the context of vehicles is the separation of the whole graph into one subgraph that contains measures that influence the vehicles characteristic properties of the vehicle itself (weight, aerodynamics, powertrain, …) and one subgraph for measures that only influence aspects outside of the vehicles properties, e.g. logistic alternatives. While the LCA-influence of a powertrain e.g. intereferes with measures that influence the energy demand of the vehicle, the transportation of the final vehicle afterwards does not influence the vehicle itself and therefore does not need to be included into the subgraph of the vehicle-properties.

6 Conclusions and Outlook

Interferences between modules within a modular LCA can cause wrong results, making it important to consider interdependencies between measures. When these possible interdependencies should be considered, the number of individual modules rises quickly to a point, where it surpasses the number of possible combinations, making the idea of modular LCA not suitable anymore. To enable the concept of modular LCA again, without losing the level of detail, that the integration of interdependencies between modules provides, two concepts of data demand reduction have been introduced. The first one is the forming of independend subgraphs for groups of modules that only influence separated properties of the product. The second one is the summarizing and excluding of the interdependency scenarios of all modules into a new additional stage, where all interdependencies for each scenario are combined into one module. With these approaches to a systematic graph design, the demand for individual modules can be reduced far below the number of possible combinations.

As the representation of the problem and the reduction of the data demand were both realized using the methods of graph theory, the following optimization of the created input data could also be performed using graph theory based algorithms like shortest path approaches, where the edges represent green house gas emissions. The shortest path through the network therefore would represent the combination of measures with the lowest level of emissions. The advantages of such an algorithm are e.g. the identification of the ideal solution, the simple elimination of technically illogical combinations by cutting out connecting edges and the possibility of expanding the optimization from single vehicles to fleet optimizations by the integration of upper boundaries on the given edges.

Alternative optimization strategies like heuristics or simulation based optimization strategies could also be applied, if the problem size becomes to big for graph theory approaches. The performance of these strategies is yet to be evaluated.

References

Broch, F.: Integration von ökologischen Lebenswegbewertungen in Fahrzeugentwicklungsprozesse, Springer Verlag (AutoUni-Publication Series, Wolfsburg) (2017)

Cerdas, F., Thiede, S., Herrmann, C.: Integrated computational life cycle engineering—application to the case of electric vehicles, CIRP Ann.–Manuf. Technol. (2018)

Domschke, W., Drexl, A., Klein, R., Scholl, A.: Einführung in Operations Research. Springer Verlag (2015)

Herrmann, I., Hausschild, M., et al.: Enabling optimization in LCA: from "ad hoc" to "structural" LCA approach—based on a biodiesel well-to-wheel case study. Int. J. Life Cycle Assess. Springer Verlag (2013)

Jungbluth, N.: Umweltfolgen des Nahrungsmittelkonsums, Öko-Institut e.V. (2000)

Koffler, C., Rohde-Brandenburger, K.: On the calculation of fuel savings through lightweight design in automotive life cycle assessments. Int. J. Life Cycle Assess. 15(1), Springer Verlag (2010)

Poppe, H.: Indikatorgestützte Umweltbewertung zur Steuerung der Produktentwicklung in der Automobilindustrie, Papierflieger Verlag (CUTEC-Publication Series 53, Clausthal-Zellerfeld) (2001)

Rohde Brandenburger, K.: Was bringen 100 kg Gewichtsreduzierung im Verbrauch?. ATZ. 115(07–08) Springer Verlag (2013)

Steubing, B., Mutel, C., Suter, F., Hellweg, S.: Streamlining scenario analysis and optimization of key choices in value chains using a modular LCA approach. Int. J. Life Cycle Assess. 21(4) Springer Verlag (2016)

United Nations: Paris Agreement, United Nations Framework Convention on Climate Change. (2015)

Vetter, P.: Leichtbau fällt nicht schwer ins Gewicht, Welt am Sonntag Nr. 49, Axel Springer SE (2017)

Integrating Life-Cycle Assessment into Automotive Manufacturing—A Review-Based Framework to Measure the Ecological Performance of Production Technologies

Malte Gebler, Felipe Cerdas, Alexander Kaluza, Roman Meininghaus and Christoph Herrmann

Abstract The transition of automotive manufacturing towards sustainability becomes more relevant when new product technologies as lightweight and electric powertrains shift environmental impacts from the use phase to the production phase. Therefore, a systemic assessment and an ecological optimization of novel production processes is necessary before implementation in factories. Furthermore, product design choices pre-determine the environmental performance of production processes. Based on a brief literature analysis of sustainable manufacturing, a framework is developed that integrates production processes with product development processes in an ecological context. The identification of ecologically-relevant core processes represents the basis for the framework development and explains, why the integration of life-cycle considerations in product development processes is decisive. Aim of the framework is to contribute to a holistic understanding of drivers that generate environmental impacts in automotive production. Furthermore, it establishes a life-cycle approach for production, which is crucial to evaluate the ecological relevance of individual resource flows to, within and from the system. The applicability of the framework is critically discussed concerning scope of the assessment, data requirements, functional unit and potential allocations problems.

M. Gebler (✉) · R. Meininghaus
Volkswagen Group Research, Environmental Affairs Production, Berliner Ring 2, 38436 Wolfsburg, Germany
e-mail: malte.gebler@volkswagen.de

F. Cerdas · A. Kaluza · C. Herrmann
Chair of Sustainable Manufacturing and Life Cycle Engineering, Institute of Machine Tools and Production Technology (IWF), Technische Universität Braunschweig, Langer Kamp 19b, 38106 Brunswick, Germany

M. Gebler · A. Kaluza · C. Herrmann
Open Hybrid LabFactory e.V., Herrmann-Münch-Str. 1, 38440 Wolfsburg, Germany

© Springer Nature Switzerland AG 2019
L. Schebek et al. (eds.), *Progress in Life Cycle Assessment*, Sustainable Production, Life Cycle Engineering and Management,
https://doi.org/10.1007/978-3-319-92237-9_6

Keywords Sustainable manufacturing · Automotive · LCA · Framework
Product-production relationship

1 Introduction

Since early the industrialization in the 19th century, industrial production has significantly accelerated the accumulation of societal welfare, but as well caused an increasing damage to the environment. The transformation of natural resources in industry to goods and services generates by-products and emissions, which impact locally and globally on the environment (Herrmann et al. 2015). Climate change, water and air pollution, resource depletion and biodiversity loss represent the main environmental impacts, which are associated with industrial activities. Recent scientific findings show that anthropogenic interference has exceeded some of the planetary boundaries (Rockström et al. 2009). Global production and consumption systems need to be "re-tuned" to avoid a strong impact on human livelihoods described as "sustainable development" by the Brundlandt Commission (WCED 1987).

In automotive production, the variety of applied processes, used materials and required to produce a conventional car is great (Schmidt et al. 2004; Rivera and Reyes-Carrillo 2016). Current developments in automotive engineering imply a shift towards electric powertrains and lightweight car bodies (Tagliaferri et al. 2016; Schmidt et al. 2004). These transformations will induce new production technologies and processes to the factories and potentially change the ecological performance of car manufacturing as especially battery production is associated with high energy demands (Romare and Dahlöf 2017; Notter et al. 2010). However, transformative processes in manufacturing are multi-dimensional, dynamic, complex and are often not comprehensively designed due to lack of understanding (Moldavska 2016). The aim of this article is to provide a systemic approach, how vehicle manufacturing processes can be systematically described and ecologically assessed. For this purpose, a framework is developed and critically discussed concerning its applicability.

2 Methodological Approach for Framework Development

The development of the framework merely focuses on the understanding of the ecological dimension of automotive manufacturing. This is due to the fact that—following an advanced understanding of sustainability by Rockström (2015)—the environmental dimension of sustainability represents the foundation for social and economic activity. The development of a framework is based on a brief literature analysis and divided into four steps (see Table 1): (1) a summary of the current state of sustainable manufacturing categories; (2) the adaptation of these categories on automotive production to identify core processes of the framework; (3) the determination of relevant elements (factors, determinants) of the core processes that determine

Table 1 Methodological approach for framework development

Step	Step 1	Step 2	Step 3	Step 4
Content	Literature analysis and identification of sustainable manufacturing categories	Identification of ecologically-relevant core processes of automotive production	Subdivision of core processes into system components	Identification of relationships among system components and framework development

system components; (4) the identification of (inter-)relationships of these system components as well as system boundaries to establish a framework.

Literature analysis to identify categories and principles of sustainable manufacturing

First, of existing approaches to and relevant aspects of sustainable manufacturing are analyzed: automotive production, sustainable manufacturing, industrial ecology, Life Cycle Engineering (LCE) and Life-Cycle Assessment (LCA). Aim of the analysis is to evaluate definitions, concepts, frameworks and methods that represent a foundation for the first purpose of this study to identify "categories" of sustainable manufacturing.

Adaptation of sustainable manufacturing on automotive production to identify core processes

Hence, the identified categories and principles of sustainable manufacturing are related to automotive production to identify "core processes" of the framework. These determine key relationships of relevant elements and layers of manufacturing for the development of a systemic understanding and for the definition of system boundaries.

Determination of system components and framework development

The previously defined core processes are divided into their relevant "system components". These represent steps, parts or evolutionary phases and can be described as information or resource flows. Finally, the logical relationships of the individual system components are identified to establish the framework. Therefore, it is crucial to understand, how individual system components are (inter-)related.

3   Results

3.1   Categories of Sustainable Manufacturing

Sustainable manufacturing represents an urgent but as well a very broad research field. As of 2017, various approaches concerning sustainable manufacturing exist.

The definition by the U.S. Department of Commerce is commonly applied (Moldavska and Welo 2017):

> [Sustainable manufacturing is the] creation of discrete manufactured products that in fulfilling their functionality over their entire life-cycle cause a manageable amount of impacts on the environment (nature and society) while delivering economic and societal value.

Despite ongoing research for decades, most scientific contributions have been identified being published since 2010 (Hartini and Ciptomulyono 2015; Moldavska and Welo 2017). Current research highlights various definitions, frameworks, measurability, metrics and methods to grasp the multi-dimensional and inter-disciplinary challenge (Haapala et al. 2013). Moldavska and Welo (2017) have conducted an extensive literature review and point out relevant categories (see Table 3) , which define sustainable manufacturing in its complexity. According to their analysis, sustainable manufacturing approaches the production of products/services from both a triple-bottom-line and life-cycle perspective. Sustainability in manufacturing should be integrated in business models (not vice versa) and is understood in a two-fold-manner—to produce in a sustainable manner and to produce sustainable products.

3.2   Core Processes of Automotive Production in the Context of Sustainable Manufacturing

The application of a life-cycle perspective on automotive manufacturing leads to the production life-cycle, which enables a description of relevant resource flows during each life-cycle phase. Applied reduction strategies as cleaner production (UNEP 2001), symbiotic use and closed loop, resource re-utilization/recovery (Chertow 2007) can be systematically assessed concerning their influence on the environmental impact. The categories "integrating perspective" as well as the "relationship between sustainability and manufacturing" can be understood as the biosphere-technosphere relationship of automotive production. Sustainability should be integrated in production processes to produce in a sustainable manner (Moldavska and Welo 2017). This relationship has been conceptually described through the concept of Industrial Ecology (IE) which aims to provide tools and methods to understand their complex relationship (Ehrenfeld 1997). The categories "domain", "potentials to enhance" and "potentials to decrease" can be related to the product-production relationship. Product design decisions have an impact on the ecological performance of production processes as a specific product design implies discrete manufacturing processes (Götze et al. 2014). The Integrated Framework for Life-Cycle Assessment (Hauschild et al. 2017) has been developed to relate product/production engineering activities to an absolute sustainability context. This implies the recognition of planetary boundaries as defined in by Rockström et al. (2009). The framework is considered useful, as it enables assessments, from both the technology level (bottom-up) and the global sustainability level (top-down).

3.3 System Components

The previously identified core processes are subdivided into system components, which are relevant for the framework development. The system components represent sub-processes and can be considered as variables that influence the ecological performance of production.

Product-production relationships

The integrated LCE-framework (Hauschild et al. 2017) is considered valid to describe the (inter-)relationship between product design and production in a factory context as it differentiates between different engineering life-cycle phases. Kaluza et al. (2016) present and apply this framework and highlight four relevant stages of automotive product development. Their distinction in "product specification", "concept development", "detailed development" and "production preparation" (Kaluza et al. 2016) will be used to describe the product/production-relationship.

Production life-cycle

Applying life-cycle thinking on production, the understanding of each life-cycle phase differs when comparing it to products (see Table 2). The first phase "raw materials" represents the material/energy requirements. The "production" phase represents the production of process material (pre-chains). The "use phase" of a product comprehends energy and material requirements for the production process in the factory. Finally, "end-of-life" represents the disposal/recycling phase, which describes the impacts of by-products such as airborne emissions, wastewater, waste or other by-products.

Biosphere-technosphere relationship

The biosphere-technosphere relationship of manufacturing is conceptually evaluated by concepts as Industrial Ecology (Ehrenfeld 1997) and Industrial Symbiosis (Chertow 2007) which describes the exchange of material and energy flows from ecosystems to industries and back. The biosphere represents a mandatory precondition for the technosphere, as it provides natural resources and represents the

Table 2 Distinction between product and production life-cycles

Life-cycle	Raw materials	Production	Use phase	End-of-life	Functional unit
Product	Product materials and energy	Manufacturing process	Energy and material requirements for utilization	Product recycling/disposal	A discrete good/service
Production	Energetic, organic and abiotic resources	Pre-chain process of process materials	Manufacturing process	Emissions Waste Waste water	A discrete manufacturing result

sink for undesired production outputs (emission, wastewater, waste). The relationship of both spheres can be described as a system boundary with "natural resources" (input) and "emissions, waste, wastewater" (output) on the biospheric side, while the transformation of these resources to pollution occurs within the technosphere.

Resource flows (input-transformation-output)

The transformation of natural resources in industrial processes to (undesired) emissions or pollutions requires a clear distinction for. Raw materials are transformed into pre-products, which are hence transformed in the production processes to by-products and might finally become an emission, a waste or concentrations in wastewater. IS enables the consideration of recovery/secondary utilization or closed loop of these flows among partners in industries to enable a secondary use of by-products (Chertow 2007). Six system components concerning natural-industrial material flows are identified as "natural resource demands", "production pre-chain (process inputs)", "production", "by-products (process outputs)", "energy/material recovery" and "emissions, waste, wastewater" (Table 3).

3.4 Identification of Relationships and Framework Development

Product development-production preparation

The product-production relationship describes an engineering process that connects product development with subsequent production planning (Kaluza et al. 2016). Product design choices pre-determine ecological impacts of the production as the product design defines the planned production machinery and process (e.g. the number of welding points). Relevant product information is therefore necessary to describe the subsequent production process comprehensively. Approaching the product development from an LCE perspective, the relationship between product and production is represented through evaluating future (ecological) impacts of the production process (Hauschild et al. 2017) (Fig. 1).

Production life-cycle, resource flows and biosphere-technosphere relationship

The production life-cycle is characterized through a transformation of natural resources into finished goods and undesired by-products. To ensure transparency within the factory, the production is divided into "unit process", "process chain" and "facility/technical building services" according to the multi-level factory perspective of Duflou et al. (2012). To supply the production with the necessary resources, resource flows from production pre-chains represent the necessary inputs. Downstream, the process generates by-products (excess energy or undesired process outputs) which represent either input flows back to the production (recovery)/to other industries (secondary use) or leave the technosphere. as airborne emissions, waste

Table 3 Categories of sustainable manufacturing and their adaptation on automotive production

Sustainable manufacturing category	Content	Automotive production	Core process	System components
Life-Cycle perspective	Total life-cycle as perspective for product assessment	Total life-cycle as perspective for production assessment	Production life-cycle	Natural resources Production pre-chain Production Emissions, waste, waste water
Time perspective	Not extensively discussed, but both short- and long-term thinking is mentioned in some publications	Short- and long-term	*Depending on case or scope*	
Integrating perspective	TPL as a concept to combine economic, environmental and social dimensions of manufacturing	Integration of sustainability in production processes (not vice versa)	Biosphere-technosphere relationship	Natural resources (input) emissions, waste, wastewater (output)
Relation between sustainability and manufacturing	Description in a two-fold manner: 1. Manufacturing for sustainability (of sustainable products) 2. Sustainability of manufacturing (produce in a sustainable manner	Sustainability of automotive production (produce in a sustainable manner)		
Domains	Product, process, community, customers, employees	Product and process (product design and production planning processes)	Product-process relationship (as information process)	Product specification Concept development Detailed development production preparation Production preparation
Potentials to enhance	Economic benefits, natural environment, safety	Integrated engineering and planning processes		
Potentials to decrease	Resources (non-specified), energy, materials, pollution (non-specified), wastes, toxic materials, pollution to air	Cleaner production, recovery/recycling, closed loop, symbiotic use	Resource flows (input-transformation-output)	Natural resource demands Pre-chain (process inputs) Production By-products (process outputs) Energy/material recovery Emissions, waste, wastewater
Moldavska and Welo (2017)		*Own adaptation*		

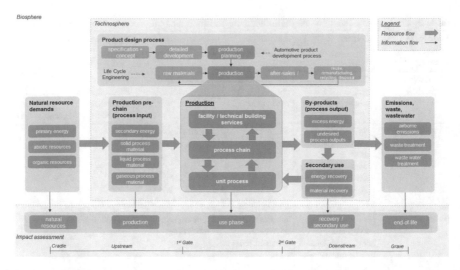

Fig. 1 Framework to measure the ecological performance of production technologies

or waste water. Therefore, the technosphere-biosphere relationship is characterized through a system boundary, whereas natural resource demands and emissions, waste and wastewater are on the biosphere side while their utilization in the production process occurs in the technosphere.

4 Discussion

The framework enables an iterative engineering process, which evaluates future ecological impacts of the manufacturing life-cycle due to a specific automotive product design. Once conducted, the derived information can be used to optimize the product life-cycle until an optimum has been reached. A detailed impact analysis of each input or output flow concerning their ecological relevance and type of impact (human health, ecosystem, resources) is possible. Depending on the scope, the production can be evaluated concerning upstream (Cradle-to-Gate), in-factory (Gate-to-Gate), downstream (Gate-to-Grave) or the entire life-cycle (Cradle-to-Grave) impacts. In Gate-to-Gate assessments, a distinction of impacts between unit process, process chain and facility/technical building services is possible. Furthermore, the frameworks enables a holistic comparative assessment of different production process variants through applying it to different process designs. Problem-shifting, as a common engineering phenomenon, can be excluded as the framework is based on life-cycle considerations.

This implies a careful and distinct definition of the scope and the functional unit as well as a thorough process understanding including up- and downstream processes. The framework enables a narrow (unit process) or a wide scope (entire factory).

Therefore, cut-offs need to be carefully selected der to avoid double counting or excluding relevant resource flows. This implies the availability of data or profound assumption, which might lead to greater uncertainties in the results. The choice and definition of the functional unit, which should relate to the desired outcome of the production, represents a critical process. Depending on the time perspective, the functional unit could represent a single process result (short-term) or the accumulation of process results of an annual production (long-term). The results might differ due to the allocation of indirect resource flows as e.g. the factory heating. Furthermore, allocation problems need to be solved, when different products are produced on one production line. Therefore, multi-functionality of production systems has to be taken into account when applying the framework on entire factories or in a long-term time perspective. The allocation of impacts could be solved through averaging (per produced item/product) or specification: per production output (produced number of specific product/total number of produced product), per required production space (area per specific product/total area), per economic considerations (revenue per specific product/total revenue) or per produced time (production time per specific product/total production time).

5   Conclusion

Automotive manufacturing represents a complex process as it requires complex process chains and applies a variety of processes to produce cars of a certain quality within a set time frame. A transformation towards ecological sustainability represents therefore a multi-dimensional process, which requires a holistic and systemic frame to include all relevant aspects that influence the ecological performance. This work focuses on the ecological dimension as it represents the basis for a sustainable socio-economic development. Based on a literature analysis of sustainable manufacturing and its categories, this paper identifies core processes of automotive production that impact on the environment. These core processes are evaluated concerning their relationships to establish a framework that enables a systematic and holistic impact assessment of automotive production processes. It furthermore enables the estimation of production impacts due to product design processes and is therefore useful to support LCE approaches in automotive engineering. Current changes in car technologies (e.g. lightweight, electric powertrains) induce changes in production processes and therewith the ecological performance of factories, too. Therefore, a thorough and holistic evaluation of these new production technologies is necessary and urgent. We propose the aim to generate an early and thorough ecological understanding of new manufacturing process before they are integrated into automotive factories.

References

Chertow, M.R.: "Uncovering" industrial symbiosys. J. Ind. Ecol. **11**, 11–30 (2007)

Duflou, J.R., Sutherland, J.W., Dornfeld, D., Herrmann, C., Jeswiet, J., Kara, S., Hauschild, M., Kellens, K.: Towards energy and resource efficient manufacturing: a processes and systems approach. CIRP Ann.—Manuf. Technol. **61**, 587–609 (2012)

Ehrenfeld, J.R.: Industrial ecology: a framework for product and process design. J. Clean. Prod. **5**, 87–95 (1997)

Götze, U., Schmidt, A., Symmank, C., Kräusel, V., Rautenstrauch, A.: Zur Analyse und Bewertung von Produkt-Prozessketten-Kombinationen der hybriden Produktion. In: Neugebauer, R.; Götze, U.; Drossel, W.-G., 2014. Energetisch-wirtschaftliche Bilanzierung – Diskussion der Ergebnisse des Spitzentechnologieclusters eniPROD, 3. Methodenband der Querschnittsarbeitsgruppe "Energetisch-wirtschaftliche Bilanzierung" des Spitzentechnologieclusters eniPROD, Wissenschaftliche Scripten, Auerbach, 2014 (2014)

Haapala, K.R., Zhao, F., Camelio, J., Sutherland, J.W., Skerlos, S.J., Dornfeld, D.A., Jawahir, I.S., Clarens, A.F., Rickli, J.L.: A review of engineering research in sustainable manufacturing. J. Manuf. Sci. Eng. **135**, 1–16 (2013)

Hartini, S., Ciptomulyono, U.: The relationship between lean and sustainable manufacturing on performance: literature review. Procedia Manuf. **4**, 38–45 (2015)

Hauschild, M.Z., Herrmann, C., Kara, S.: An integrated framework for life-cycle engineering. Procedia CIRP **61**, 2–9 (2017)

Herrmann, C., Blume, S., Kurle, D., Schmidt, C., Thiede, S.: The positive impact factory—transition from eco-efficiency to eco-effectiveness effectiveness strategies in manufacturing. Procedia CIRP **29**, 19–27 (2015)

Kaluza, A., Kleemann, S., Broch, F., Herrmann, C., Vietor, T.: Analyzing decision-making in automotive design towards life cycle engineering for hybrid lightweight components. Procedia CIRP **50**, 825–830 (2016)

Moldavska, A.: Model-based sustainability assessment—an enabler for transition to sustainable manufacturing. Procedia CIRP **48**, 413–418 (2016)

Moldavska, A., Welo, T.: The concept of sustainable manufacturing and its definitions: a content-analysis based literature review. J. Clean. Prod. **144**, 744–755 (2017)

Notter, D.A., Gauch, M., Widmer, R., Wäger, P., Stamp, A., Zah, R., Althaus, H.-J.: Contribution of Li-ion batteries to the environmental impact of electric vehicles. Environ. Sci. Technol. **44**, 6550–6556 (2010)

Rivera, J.L., Reyes-Carillo, T.: A life cycle assessment framework for the evaluation of automobile paint shops. J. Clean. Prod. **155**, 75–87 (2016)

Romare, M., Dahlöf, L.: The Life Cycle Energy Consumption and Greenhouse Gas Emissions from Lithium-Ion Batteries—A Study with Focus on Current Technology and Batteries for Light-Duty Vehicles. Report Number: C 243. The Swedish Environmental Institute, Sweden (2017)

Rockström, J., Steffen, W., Noone, K., Persson, Å., Chapin III, F.S., Lambin, E.F., Lenton, T.M., Scheffer, M., Folke, C., Schellnhuber, H.J., Nykvist, B., de Wit, C.A., Hughes, T., van der Leeuw, S., Rodhe, H., Sörlin, S., Snyder, P.K., Costanza, R., Svedin, U., Falkenmark, M., Karlberg, L., Corell, R.W., Fabry, V.J., Hansen, J., Walker, B., Liverman, D., Richardson, K., Crutzen, P., Foley, J.A.: A safe operating space for humanity. Nature **461**, 472–475 (2009)

Rockström, J.: Bounding the Planetary Future: Why We Need a Great Transition. The Great Transition Initiative. http://www.greattransition.org/publication/bounding-the-planetary-future-why-we-need-a-great-transition (2015)

Schmidt, W.-P., Dahlqvist, E., Finkbeiner, M., Krinke, S., Lazzari, S., Oschmann, D., Pichon, S., Thiel, C.: Life cycle assessment of lightweight and end-of-life scenarios for generic compact class passenger vehicles. Int. J. Life-Cycle Manage. **9**, 405–416 (2004)

Tagliaferri, C., Evangelisti, S., Acconcia, F., Domenech, T., Ekins, P., Barletta, D., Lettieri, P.: Life cycle assessment of future electric and hybrid vehicles: a cradle-to-grave systems engineerings approach. Chem. Eng. Res. Des. **112**, 298–309 (2016)

UNEP: International Declaration on Cleaner Production: Implementation Guidelines for Facilitating Organizations. United Nations Environment Programme, Nairobi/Kenia (2001)

WCED: Our Common Future—The World Commission on Environment and Development. Oxford University Press, United Kingdom (1987)

Part III
Product System and Inventory Modelling

Hydrothermal Carbonization (HTC) of Sewage Sludge: GHG Emissions of Various Hydrochar Applications

Fabian Gievers, Achim Loewen and Michael Nelles

Abstract Sewage sludge contains valuable nutrients like phosphorus (P) as well as a whole series of harmful substances. Therefore, conditioning should be designed to remove those pollutants. In Germany sewage sludge is treated mainly at thermal facilities such as sewage sludge mono-incineration plants, cement plants or coal fired power plants. However, ecological impacts of new treatment methods like hydrothermal carbonization (HTC) remain unknown. In the study presented in this paper, the complete life cycles of the carbonization process of sewage sludge (5% dry matter) with associated auxiliary flows (e.g. electricity and naturals gas) and different applications of the produced char were modelled. In order to identify the environmentally most promising and sustainable application, four different scenarios of hydrochar utilization as fuel or fertilizer were analyzed. The resulting global warming potentials (GWP) after ReCiPe midpoint methodology were calculated. Results show that the best scenario in environmental terms has savings of 0.074 kg CO_2 cq/kg. The highest emissions were observed for the agricultural use of hydrochar as a substitute for NPK-fertilizer with 0.025 kg CO_2 eq/kg, which even outnumbers the emissions of the benchmark process chain of sewage sludge mono-incineration (0.013 kg CO_2 eq/kg). Results underline the sustainability of hydrothermal carbonization of sewage sludge as compared to sewage sludge mono-incineration.

Keywords Life cycle assessment · Hydrothermal carbonization · Sewage sludge Hydrochar · Global warming potential · Sustainable sludge management

F. Gievers (✉) · A. Loewen
Faculty of Resource Management, HAWK-University of Applied Sciences and Arts,
Rudolf-Diesel-Straße 12, 37075 Göttingen, Germany
e-mail: fabian.gievers@hawk.de

F. Gievers · M. Nelles
Faculty of Waste and Resource Management, University of Rostock, Justus-von-Liebig-Weg 6,
18059 Rostock, Germany

M. Nelles
DBFZ-German Biomass Research Centre, Torgauer Str. 116, 04347 Leipzig, Germany

© Springer Nature Switzerland AG 2019
L. Schebek et al. (eds.), *Progress in Life Cycle Assessment*, Sustainable
Production, Life Cycle Engineering and Management,
https://doi.org/10.1007/978-3-319-92237-9_7

1   Introduction

The growing demand for fertilizers due to a growing world population increases the pressure on limited natural nutrient resources. In particular, the supply routes of phosphorus, a non-renewable resource that could be depleted in 50 to 100 years, need to be reconsidered (Cordell et al. 2009; Sartorius et al. 2011). Recycling of materials with high phosphorus content must therefore be improved in order to secure the livelihoods of future generations. Sewage sludge is an important natural and locally available phosphorus source (Cordell and White 2011; Schoumans et al. 2015; Klinglmair et al. 2015). Existing treatment methods such as co-combustion in the cement industry or co-incineration in lignite-fired power plants reduce the P concentration in the ash and thus make phosphorus extraction more difficult and expensive. An ideal technology should offer maximum P-recovery rates, removal and destruction of potentially hazardous substances such as heavy metals, organic micropollutants and pathogens, good fertilising properties of the product, a good profitability and low environmental risks. (Egle et al. 2016; Leinweber et al. 2018). Therefore the Hydrothermal Carbonization (HTC) of sewage sludge is investigated as a promising approach for implementing a circular economy for nutrients (Brookman et al. 2016; Heilmann et al. 2014; Zhao et al. 2017) and a sustainable energy generation (Stucki et al. 2015; Libra et al. 2011; Titirici et al. 2007) with simultaneous decrease of pathogens and other organic pollutants (Vom Eyser et al. 2015, 2016; Weiner et al. 2013). HTC is a thermochemical process in aqueous phase under saturated pressure and temperatures between 160 and 250 °C. Typically, over several hours biomass is converted into a valuable solid coal (hydrochar), partially dissolved fractions in the aqueous phase and a small amount of gases (usually CO_2) (Berge et al. 2011; Bergius 1932; Kruse et al. 2013; Funke and Ziegler 2010). In recent years, there has been a growing interest in industrial applications of HTC as a waste treatment method and in the usability of the produced hydrochar (Hoekman et al. 2013; Buttmann 2011; Stucki et al. 2015). To evaluate the sustainability of this new approach, life cycle assessment (LCA) studies of HTC of different biomass feedstocks were performed (Owsianiak et al. 2016; Benavente et al. 2017; Liu et al. 2017; Stucki et al. 2015; Berge et al. 2015). In this study, an LCA of HTC of sewage sludge digestate and four different hydrochar applications were carried out an the results were compared to the usual process chain of sewage sludge mono-incineration with subsequent ash landfilling. The main goal was to identify the best utilization for hydrochar from sewage sludge characterised by the lowest CO_2 footprint in comparison to the benchmark process of mono-incineration.

2 Methods

2.1 Material and Energy Flows

A model was set up for the HTC process and used to examine the carbonization of sewage sludge with a dry matter (DM) content of 5% (organic solids content in dry matter (oDM): 48%) after anaerobic digestion in an existing wastewater treatment plant (WWTP) (Table 1).

Hydrothermal Carbonization Parameters:

The process parameters of the HTC were set to the following values in the model (Table 1):

Material flows:

Every sewage sludge comes with a certain load of inorganics and heavy metals. While organic compounds react during the process, heavy metals cannot be destroyed and accumulate in the solid fraction (Yue et al. 2016). Since their accumulation has a toxic risk potential, the concentrations of heavy metals have to be carefully observed (Libra et al. 2011). Beside inorganic hazardous substance sewage sludge contain organic pollutants such as polychlorinated dibenzo-dioxins (PCDD) and polychlorinated dibenzo-furans (PCDF), polychlorinated biphenyl (PCB), pharmaceuticals or pesticides. These substances can be degraded or reformed, regenerated and accumulated in certain fractions through hydrothermal process conditions (Weiner et al. 2013; Tirler and Basso 2013). Therefore, hydrothermal carbonsiation is a good way to treat sewage sludge in order to destroy harmfull organics while saving the availability of nutrients, such as N, K and P. In this study flows and accumulation of heavy metals, organic pollutants and valuable nutrients were modelled to identify the best utilization of hydrochar from sewage sludge. Carbonization of sewage sluge with very high amounts of water mainly results in process water, which contains soluable

Table 1 Model parameters for sewage sludge and carbonization conditions

Parameter	Unit	Value
Sewage Sludge		
Dry matter (DM)	(%)	5
Organic dry matter (DM)	(% of DM)	48
Specific heat capacity	(kJ kg^{-1} K^{-1})	4.58
Sludge input temperature	(°C)	38
Hydrothermal Carbonization		
Carbonization temperature	(°C)	220
Duration	(h)	4

organics, nutrients and inorganics. In the model, the treatment of the generated process liquid in a WWTP, characterized through investigated concentrations of C, N and P, was implemented.

Energy flows:

The thermochemical modelling of the energy flows of the carbonization process was carried out with a mathematical modelling approach. Initially, the specific heat capacities of the individual parts of the sewage sludge (water, organic and inorganic substances) were used to determine the energy requirements for the carbonization process. The results were confirmed with literature data. Other energy flows, such as the electricity required for the dewatering process, were also taken from the literature. The aggregated processes of system extension to include emission credits have been taken from the above-mentioned databases. The data for the benchmark process of mono-incineration was taken from generic data by econinvent.

2.2   Life Cycle Assessment

The LCA was conducted in accordance with the requirements of the ISO standard 14044:2006 (DIN EN ISO 14044 2006).

Functional Unit:

The main function of the hydrothermal carbonization of sewage sludge is to stabilise the sludge produced. In order to compare the HTC of sewage sludge with mono-incineration, the functional unit was defined as follows:"Treatment of 1 kg of sewage sludge from anaerobic digestion with a dry matter content of 5%".

System Boundaries:

The boundary of the system includes the carbonization of digested sewage sludge, possible transportation and storage, power and heat generation and char application. For comparison purposes the benchmark process of sewage sludge mono-incineration was also investigated. The examined carbonization of sewage sludge covered the construction and decommissioning of the HTC-plant, the actual carbonization process of the sludge, linked energy and equipment provisions as well as the separation, storage and transportation of hydrochar by truck to the respective location considered in the four different scenarios. The energy for the HTC and the filter press for separating hydrochar from process water were provided as electricity for pumping the sludge and natural gas combustion for providing the necessary process-heat. Treatment of HTC process water was performed in a WWTP on side, characterized by C, N and P-content. The utilization of hydrochar as fuel or fertilizer was compared to combustion and application of fossil-based products: NPK-fertilizer, peat, municipal solid waste (MSW) and lignite. Therefore, the avoided burden approach was performed to consider the emissions of processes replaced by the HTC process chains. Substitution

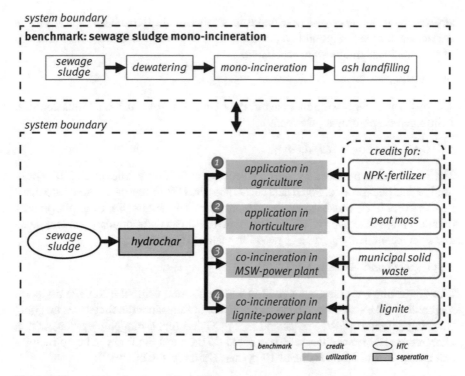

Fig. 1 Different ulitizations scenarios of hydrochar with system boundaries of process chains and benchmark scenario

of lignite and municipal solid waste (MSW) was balanced by the energy content of the hydrochar for co-combustion in power plants using existing incineration capacities. For avoided NPK-fertilizer application the nutrients content of hydrochar was taken into account. Credits for peat in horticulture were calculated as substitution by weight. Altogether, four different utilization paths were analyzed and compared regarding emissions of appropriate benchmark processes (Fig. 1).

Geographic scope:

Although sustainable nutrient recycling is not bound to a specific region, Germany served as a model region for the assessment. All data concerning hydrochar and sewage sludge characterization were taken from different former studies located in Germany. The background processes such as electricity mix or natural gas supply, which were taken from the database of GaBi, were also based on german backround processes. If background data were not available for Germany, either European or Swiss data were used.

Modeling Framework:

Since HTC is not yet a market-penetrating technology for treating sewage sludge, and the production and use of hydrochar as a fuel or substrate is unlikely to result in any structural changes in the near future, an attributional approach was applied in the assessment for the foreground system of HTC. In cases of processes with substitution of commodities, credits were accounted.

LCA-software:

The product systems were modeled in the LCA software GaBi 8.1 (thinkstep AG, Leinfelden-Echterdingen, Germany).

Life Cycle Inventory (LCI) data:

The LCI datasets provided by GaBi and ecoinvent (v3.3) (Wernet et al. 2016) were used as data background. Some data concerning the HTC-Process and some auxiliary flows were either estimated, calculated or taken from literature. For example, process parameters of HTC-plant were first determined based on thermodynamic calculations and then reconciled with data of two pilot scale HTC plants.

Life Cycle Impact Assessment (LCIA):

Life Cycle Impact Assessment (LCIA) was performed using the ReCiPe midpoint methodlogy 2016, as implemented in GaBi TS. In this paper the focus was on gobal warming potential (GWP) measured in kg CO_2 eq (excl. biogenic carbon) with a Hierarchist (H) perspective, which is based on the most common policy principles and uses a medium time frame of 100 years (Huijbregts et al. 2017).

Assumptions and limitations:

Due to the lack of data from HTC plants and hydrochar users on an industrial scale, the material flows (including transport routes and the weighting and use of modifications from generic data) were determined on the basis of reasonable assumptions and data from pilot plants and literature. In addition, the geographical scope only includes energy data from Germany and there were only four scenarios modelled for the use of hydrochar.

3 Results and Discussion

Benchmark Process:

The benchmark process of sewage sludge digestate mono-incineration was based on a process from the generic ecoinvent database: [Jungbluth, N., treatment of digester sludge, municipal incineration, future, CH, Substitution, consequential, long-term, ecoinvent database version 3.3]. The functional unit of the process refers to the mono-incineration of 1 kg of wet sludge with 95% water and is therefore comparable to the results of the HTC model. For the mono-combustion of sewage sludge, 0.013 kg CO_2-equivalent/kg of sewage sludge were determined as benchmark emissions. In this process, all relevant flows were considered, in particular the dewatering of the sludge

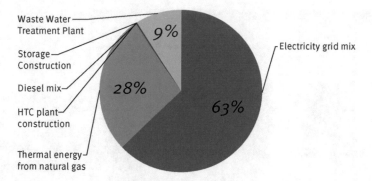

Fig. 2 Distribution of GHG emissions for HTC process

and all auxiliary flows associated with the sludge treatment. Credits were giving for energetic output of the incineration process (electricity mix of Germany). As the current state of commercial HTC has not reached market penetration, a process with future energy demands and emission reductions has been chosen for the benchmark process.

Emissions associated with HTC:

The distribution of emissions for the entire HTC process was analyzed to identify the processes with the highest environmental impact (Fig. 2). The generation of electricity for running the HTC plant and auxiliary processes lead to the main part of total CO_2 eq emissions (63%). In addition to power demand, the heat supply by natural gas resulted in 28% of the emissions. These results underline the importance of the water content of sewage sludge for optimizing the environmental performance of HTC plants. The energy consumption for heating up the carbonization process and the energy for pumping increases with the water content of the feedstock (Owsianiak et al. 2016). Furthermore, a broad spectrum of energy consumption of dewatering technologies can be observed, also for the same technology (Yoshida et al. 2013). Therefore, further investigations in large-scale HTC plants should determine the optimal process parameters for dewatering before or after the carbonization process. One option would be the firing of HTC reactors with biogas from the wastewater treatment plant to replace natural gas. A further energetic optimisation of the HTC would be to increase the dry matter content of sewage sludge before carbonization. The third relevant source of CO_2 eq emissions of HTC is the treatment of the process water (9%). By changing the process water treatment from aerobic to anaerobic, emissions could be further reduced (Wirth et al. 2012). Overall, the treatment of sewage sludge with HTC resulted in emissions of 0.051 kg CO_2 equivalent per kg of sewage sludge (Fig. 3).

Hydrochar utilization scenarios:

The net emissions derived from HTC, application of the hydrochar and credits are distributed as follows (Fig. 3): The first scenario of agricultural use of hydrochar has

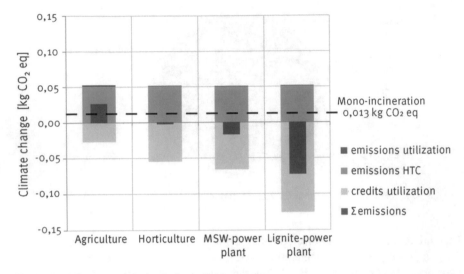

Fig. 3 Greenhousegas emissions of examined scenarios

total emissions of 0.025 kg CO_2 equivalent per kg of sewage sludge. The credits are the lowest of all scenarios due to the relatively low nitrogen content of hydrochars. By substituting NPK fertilizers in agriculture, the greatest greenhouse gas emissions could be saved by the avoided production of artificial nitrogen fertilizers. In addition, the agricultural use of hydrochar causes the highest emissions of all application-related CO_2 emissions. However, in comparision to the HTC emissions and the credits these emissions are relatively low and mainly derive from transportation, storage and handling activities. The second scenario leads to a saving of 0.0024 kg CO_2 equivalent per kg of sewage sludge, mainly from credits for the substitution of peat in commercial horticulture. For the third scenario of co-firing of hydrochar in waste incineration overall savings of 0.015 kg CO_2 eq per kg of sewage sludge were observed. These savings mainly depend on the substitution of fossil-based fractions of municipal solid waste such as plastics. Finally, co-firing of hydrochar in lignite power plants leads to the highest savings of 0.074 kg CO_2 eq per kg of sewage sludge.

4   Conclusion

The LCA results of HTC of sewage sludge showed that substituting fossil based fuel (lignite and parts of MSW) with hydrochar had the highest potential to reduce global warming potential (GWP) of sewage sludge treatment. In comparison to the benchmark process of mono-incineration, it is even possible to achieve negative emissions. In general, the use of hydrochar for energy purposes had a higher GHG saving potential than their use in agriculture or horticulture. An improvement in the process design of sludge treatment with HTC would be the implementation of phosphorus extraction, which could enable phosphorus recycling even in the case of energetic use of hydrochar and therefore be helpful to boost the hydrothermal carbonization of sewage sludge. However, HTC process optimization should also take other LCIA categories into account. In addition, improvements in LCIA methodology are needed in order to evaluate benefits of the sludge treatment process, such as pathogen reduction and recycling of organic matter and nutrients back to agricultural or horticultural soil. Overall, the results show that there is a more sustainable alternative to the mono-incineration of sewage sludge.

Acknowledgements This publication is a result of a project funded with federal state resources from the "Niedersächsisches Vorab".

References

Benavente, V., Fullana, A., Berge, N.D.: Life cycle analysis of hydrothermal carbonization of olive mill waste: comparison with current management approaches. J. Clean. Prod. **142**, 2637–2648 (2017)

Berge, N.D., Li, L., Flora, J.R.V., Ro, K.S.: Assessing the environmental impact of energy production from hydrochar generated via hydrothermal carbonization of food wastes. *Waste Manag. (New York, N.Y.)* **43**, 203–217 (2015)

Berge, N.D., Ro, K.S., Mao, J., Flora, J.R.V., Chappell, M.A., Bae, S.: Hydrothermal carbonization of municipal waste streams. Environ. Sci. Technol. **45**(13), 5696–5703 (2011)

Bergius, F.: Chemical Reactions Under High Pressure—Nobel Lecture (1932)

Brookman, H., Gievers, F., Loewen, A., Kirsten, Loewe: Entschärfung regionaler Nährstoffüberschüsse in Form von Gärresten und Güllen durch Anwendung der HTC, technical report. Ministry of Food, Agriculture and Consumer Protection, Lower Saxony (ML), Hannover (2016)

Buttmann, M.: Klimafreundliche Kohle durch Hydrothermale Karbonisierung von Biomasse. Chem. Ing. Tec. **83**(11), 1890–1896 (2011)

Cordell, D., Drangert, J.-O., White, S.: The story of phosphorus: global food security and food for thought. Glob. Environ. Change **19**(2), 292–305 (2009)

Cordell, D., White, S.: Peak phosphorus: clarifying the key issues of a vigorous debate about long-term phosphorus security. Sustainability **3**(12), 2027–2049 (2011)

DIN EN ISO 14044 (2006) 14044 Umweltmanagement – Ökobilanz – Anforderungen und Anleitungen. ISO (International Organization for Standardization) (2006)

Egle, L., Rechberger, H., Krampe, J., Zessner, M.: Phosphorus recovery from municipal wastewater: an integrated comparative technological, environmental and economic assessment of P recovery technologies. Sci. Total Environ. **571**, 522–542 (2016)

Funke, A., Ziegler, F.: Hydrothermal carbonization of biomass: a summary and discussion of chemical mechanisms for process engineering. Biofuels, Bioprod. Biorefin. **4**(2), 160–177 (2010)

Heilmann, S.M., Molde, J.S., Timler, J.G., Wood, B.M., Mikula, A.L., Vozhdayev, G.V., Colosky, E.C., Spokas, K.A., Valentas, K.J.: Phosphorus reclamation through hydrothermal carbonization of animal manures. Environ. Sci. Technol. **48**(17), 10323–10329 (2014)

Hoekman, S.K., Broch, A., Robbins, C., Zielinska, B., Felix, L.: Hydrothermal carbonization (HTC) of selected woody and herbaceous biomass feedstocks. Biomass Convers. Biorefinery **3**(2), 113–126 (2013)

Huijbregts, M.A.J., Steinmann, Z.J.N., Elshout, P.M.F., Stam, G., Verones, F., Vieira, M., Zijp, M., Hollander, A., van Zelm, R.: ReCiPe2016: A harmonised life cycle impact assessment method at midpoint and endpoint level. Int. J. Life Cycle Assess. **22**(2), 138–147 (2017)

Klinglmair, M., Lemming, C., Jensen, L.S., Rechberger, H., Astrup, T.F., Scheutz, C.: Phosphorus in Denmark: national and regional anthropogenic flows. Resour. Conserv. Recycl. **105**, 311–324 (2015)

Kruse, A., Funke, A., Titirici, M.-M.: Hydrothermal conversion of biomass to fuels and energetic materials. Curr. Opin. Chem. Biol. **17**(3), 515–521 (2013)

Leinweber, P., Bathmann, U., Buczko, U., Douhaire, C., Eichler-Löbermann, B., Frossard, E., Ekardt, F., Jarvie, H., Krämer, I., Kabbe, C., Lennartz, B., Mellander, P.-E., Nausch, G., Ohtake, H. and Tränckner, J.: Handling the phosphorus paradox in agriculture and natural ecosystems: scarcity, necessity, and burden of P. Ambio **47**(Suppl 1), 3–19 (2018)

Libra, J.A., Ro, K.S., Kammann, C., Funke, A., Berge, N.D., Neubauer, Y., Titirici, M.-M., Führer, C., Bens, O., Kern, J., Emmerich, K.-H.: Hydrothermal carbonization of biomass residuals: a comparative review of the chemistry, processes and applications of wet and dry pyrolysis. Biofuels **2**(1), 71–106 (2011)

Liu, X.V., Hoekman, S.K., Farthing, W., Felix, L.: Life cycle analysis of co-formed coal fines and hydrochar produced in twin-screw extruder (TSE). Environ. Prog. Sustai. Energy **36**(3), 668–676 (2017)

Owsianiak, M., Ryberg, M.W., Renz, M., Hitzl, M. and Hauschild, M.Z.: Environmental performance of hydrothermal carbonization of four wet biomass waste streams at industry-relevant scales. ACS Sustain. Chem. Eng. (2016)

Sartorius, C., Horn, J. and Tettenborn, F.: Phosphorus recovery from wastewater—state-of-the-art and future potential. In: Proceedings of the Water Environment Federation, 2011 (2011)

Schoumans, O.F., Bouraoui, F., Kabbe, C., Oenema, O., van Dijk, K.C.: Phosphorus management in Europe in a changing world. Ambio **44**(Suppl 2), S180–92 (2015)

Stucki, M., Eymann, L., Gerner, G., Krebs, R., Hartmann, F. and Wanner, R.: Hydrothermal carbonization of sewage sludge on industrial scale: energy efficiency, environmental effects and combustion (2015)

Tirler, W., Basso, A.: Resembling a "natural formation pattern" of chlorinated dibenzo-p-dioxins by varying the experimental conditions of hydrothermal carbonization. Chemosphere **93**(8), 1464–1470 (2013)

Titirici, M.-M., Thomas, A., Antonietti, M.: Back in the black: hydrothermal carbonization of plant material as an efficient chemical process to treat the CO_2 problem? New J. Chem. **31**(6), 787 (2007)

Vom Eyser, C., Palmu, K., Otterpohl, R., Schmidt, T.C., Tuerk, J.: Determination of pharmaceuticals in sewage sludge and biochar from hydrothermal carbonization using different quantification approaches and matrix effect studies. Anal. Bioanal. Chem. **407**(3), 821–830 (2015)

Vom Eyser, C., Schmidt, T.C., Tuerk, J.: Fate and behaviour of diclofenac during hydrothermal carbonization. Chemosphere **153**, 280–286 (2016)

Weiner, B., Baskyr, I., Poerschmann, J., Kopinke, F.-D.: Potential of the hydrothermal carbonization process for the degradation of organic pollutants. Chemosphere **92**(6), 674–680 (2013)

Wernet, G., Bauer, C., Steubing, B., Reinhard, J., Moreno-Ruiz, E., Weidema, B.: The ecoinvent database version 3 (part I): overview and methodology. Int. J. Life Cycle Assess. **21**(9), 1218–1230 (2016)

Wirth, B., Mumme, J., Erlach, B.: Anaerobic Treatment of Waste Water Derived from Hydrothermal Carbonization (2012). Accessed 19 July 2016

Yoshida, H., Christensen, T.H., Scheutz, C.: Life cycle assessment of sewage sludge management: a review. Waste Manage. Res.: J. Int. Solid Wastes and Public Cleansing Assoc. ISWA **31**(11), 1083–1101 (2013)

Yue, Y., Yao, Y., Lin, Q., Li, G. and Zhao, X.: The change of heavy metals fractions during hydrochar decomposition in soils amended with different municipal sewage sludge hydrochars. J. Soils and Sediments (2016)

Zhao, X., Becker, G.C., Faweya, N., Rodriguez Correa, C., Yang, S., Xie, X., Kruse, A.: Fertilizer and activated carbon production by hydrothermal carbonization of digestate. Biomass Convers. Biorefinery **19**, 292 (2017)

Uncertainty Information in LCI-Databases and Its Propagation Through an LCA Model

Alexandra Opitz and Christof Menzel

Abstract This article deals with uncertainties in particular with the uncertainty of inventory data analysis. Uncertainties cannot be avoided in LCA studies. Therefore, they should be analysed and interpreted. One problem is that in many LCA studies uncertainties are not noted. In ecoinvent, which is an example for an inventory database, the lognormal distribution is choosen as the dataset's standard distribution type. One reason is that many quantities found in nature can only take positive values. In most cases, however, the normal distribution can as well be taken as the default distribution type. In this article the uncertainty information of the ecoinvent datasets is explained and uncertainty analysing methods like the Monte Carlo simulation, der pedigree matrix or the sensitivity analysis are described. Furthermore, the convolution is mentioned as a method for analyzing the sucsessive uncertainty propagation through an LCA model. For using the convolution, the LCA data should be independent, continuous and normal distributed. In addition to that the LCA model should be a linear and not a complex system. Since the convolution is a new approach further research will be required.

Keywords LCA · Uncertainties · Uncertainty propagation · Monte carlo simulation · Pedigree matrix · Ecoinvent · Lognormal distribution · Normal distribution · Convolution

1 Introduction

Thematic origin of the research was a bachelor thesis in which the environmental impact of 48 g protein out of tofu and pork was compared. Many LCI databases (in particular ecoinvent) data were used. Furthermore, several scenarios, for example best case and worst-case scenarios, were modelled. The environmental impact results

A. Opitz (✉) · C. Menzel
Hochschule Niederrhein, Oecotrophologie, Rheydter Straße 277,
41065 Mönchengladbach, Germany
e-mail: Alexandra.Opitz@gmx.de

© Springer Nature Switzerland AG 2019
L. Schebek et al. (eds.), *Progress in Life Cycle Assessment*, Sustainable
Production, Life Cycle Engineering and Management,
https://doi.org/10.1007/978-3-319-92237-9_8

varied depending on the model. With regard to these results, the main research during the subsequent master studies dealt with the structure and uncertainty information in ecoinvent and the uncertainty propagation and analysis methods in life cycle assessment (Opitz 2016).

The following paper provides an overview of the uncertainty problem in LCA models. The focus is on the uncertainty information in LCI databases, e.g. ecoinvent, in general and its inclusion in LCA results. At the end of the paper a well-known method for determining the probability densitiy of a sum of random variables—the convolution—is described. The advantage of this method is that the probability densitiy function of an impact assessment factor can be calculated exactly, if the probability density function of each elementary and product flow is known. At the aim of using the convolution is that the uncertainty propagation through an LCA model can be analysed step by step and the uncertainty hotspots can be identified.

2 Uncertainties in LCA

Three types of uncertainties affect the outcome of an LCA model. First, there is the variation of input and output data, which can be characterized by probability distributions. Second, there is the choice of a method for calculation. Third, there is the choice of a scenario. The latter two are influenced by the author of the LCA case study. The uncertainty due to the variation of the inventory data originates from the author of the data set and can be found in the database entry (Huijbregts et al. 2004; Hauschild and Huijbregts 2015).

The focus of our work is on the uncertainty of inventory analysis data, which can be characterized by probability distributions. The main question is: "At which point do the uncertainties mainly enter the LCA model, and how do they propagate through the model?" This focus was chosen for two reasons:

(1) In many LCA case studies, the variation of inventory data is not mentioned in the interpretation phase and consequently not evaluated.
(2) The distribution type of the probability distribution for inventory data is merely a presumption.

Uncertainties cannot be avoided. Therefore, they should always be taken into consideration and interpreted in a LCA study. The interpretation of uncertainties is crucial to comparative studies as well as to models which serve as an essential basis for decisions (DIN EN ISO 14044:2006–2010; DIN EN ISO 14040:2009–2011).

3 Uncertainty Information in LCI-Databases

Uncertainty information of LCI databases is illustrated with the ecoinvent dataset "wheat grain, feed production, organic [CH]". The uncertainty information to be found in the ecoinvent database is shown in Fig. 1.

The details and terms are explained in the following paragraph:

Lognormal: This characterizes the probability density type of a dataset (here: lognormal distribution). Datasets can also have other distribution types, for instance normal or triangular distribution.

Geometric mean (μ^*): This is a parameter of the lognormal distribution. Here it describes the deterministic value of the output material "wastewater, average".

Variance of log-transformed data ($\sigma^2_{(oPedMa)}$): This is a parameter of the lognormal distribution. It is the estimated or calculated variance of the output, also called *basic uncertainty*. In case of ecoinvent, the estimated variance can be found in the table "Default basic uncertainty (variance σ^2_b of the log transformed data, i.e. the underlying normal distribution) applied to intermediate and elementary exchanges when no sampled data are available; c: combustion emissions; p: process emissions; a: agricultural emissions", which is published in Weidema et al. (2013), p. 75.

Arithmetic mean of log-transformed data (μ): This entry describes the arithmetic mean of the log-transformed data and is calculated by Eq. (1). ln () means natural logarithm.

$$\mu = \ln(\mu^*) \tag{1}$$

Variance of data with pedigree ($\sigma^2_{(wPedMa)}$): This parameter characterizes the complete variance of the output. It contains the variances of the corresponding pedigree matrix values (here 3|3|3|3|4) and the basic uncertainty. The former is also called *additional uncertainty*.

Standard deviation ($\sigma^{*2}_{(oPedMa)}$): It describes the half range of the confidence interval (without values of the pedigree matrix). It is defined by Eq. (2). The constant e means Euler's number.

Reference Products				
+ wheat grain, feed, organic	1	kg		
By-product/Waste				
− wastewater, average	3.8E-05	m3	Lognormal	1.4918
Uncertainty	Lognormal (Geometric mean=3.8E-05, Variance of log-transformed data=0.04, Arithmetic mean of log-transformed data=-10,18, Standard deviation=1.4918, CI/2wP, half range of confidence interval=1.7897, Variance of data with pedigree=0.0847) Pedigree matrix: 3 3 3 3 4			

Fig. 1 Illustration of uncertainty information of an ecoinvent dataset ("wheat grain, feed production, organic [CH]"). In the category "uncertainty" we read: Uncertainty: Lognormal (Geometric mean = 3.8E-05, Variance of log-transformed data = 0.04, Arithmetic mean of log-transformed data = −10, 18, Standard deviation = 1.4918, CI/2wP, half range of confidence interval = 1.7897, Variance of data with pedigree = 0.0847) Pedigree matrix: 3 3 3 3 4 (Weidema et al. 2013)

$$\sigma^{*2}_{(oPedMa)} = e^{\left(\sqrt{basic\ uncertainity}\right)^2} = e^{\left(\sqrt{\sigma^2_{oPedMa}}\right)^2} \tag{2}$$

<u>CI/2wP, half range of confidence interval</u> ($\sigma^{*2}_{(mPedMa)}$): This parameter describes the half range of confidence interval (the values of the pedigree matrix are now included). It is determined by Eq. (3).

$$\sigma^{*2}_{(wPedMa)} = e^{\left(\sqrt{basic + additional\ uncertainty}\right)^2} = e^{\left(\sqrt{\sigma^2_{wPedMa}}\right)^2} \tag{3}$$

In the example mentioned the lognormal distribution is adopted as distribution type. An analysis of a number of ecoinvent datasets reveals that many datasets are assumed to be lognormally distributed. This distribution type, however, is quite frequently not deduced from data.

In the literature, the following reasons are stated to justify the assumption that the lognormal distribution is the standard distribution type for ecoinvent datasets (when not enough deterministic values are available):

Many quantities found in nature can take positive values only. Hence, they can be characterized by a right-skewed and asymmetric distribution. In addition they are multiplicative instead of additive, which due to the central limit theorem results in a lognormal distribution (Weidema et al. 2013; Koch 1966; Bourgault 2016; Limpert et al. 2001a, b).

The lognormal distribution as standard distribution type is the simplest way to describe uncertainties in datasets that are based on a few data only, because of the use of an estimated variance as *basic uncertainty* plus the pedigree matrix. The values of the pedigree matrix are also lognormally distributed datasets. They have the value zero as geometric mean and the value mentioned within the pedigree matrix as standard deviation (Weidema et al. 2013; Muller et al. 2016; Mutel 2013; Suh et al. 2016).

Negative values cannot be generated during a Monte Carlo Simulation (which is described in Sect. 4), since a lognormal distribution takes positive values only (Bourgault 2016; Feck 2007; Publication Office of the European Union 2010).

A study of Qin and Suh (2016) shows, that many datasets follow a lognormal distribution rather than other distribution types (Qin and Suh 2016).

Although there are many reasons that support the lognormal distribution as standard distribution type, the normal distribution could as well be taken as the default one. If the normal distribution is the standard distribution type, statistical methods can be applied easier and uncertainty propagation through an LCA model can as well be calculated in an unsophisticated way (see Sect. 5).

Depending on the position of the expectation value of the normal distribution, the probability of obtaining a negative value for the variable in question tends towards zero. In addition, further developments of the pedigree matrix allow introducing other distribution types (Muller et al. 2016).

Many datasets are based on a few, or even one, sample. Because of this, the dataset's variance, the *variance of data with pedigree,* is estimated with the *basic uncertainty* and the pedigree matrix. In addition, lognormally distributed elements take positive values only. Therefore, it is assumed that the lognormal distribution is arbitrarily chosen as standard distribution type. Nevertheless if there is no detailed information about the dataset's original distribution type, the lognormal distribution may not be the best way to describe the dataset's variability (Bourgault 2017; Hedderich and Sachs 2015).

In the sequel, the normal distribution is chosen as standard probability density function type for methodological and technological simplifications. In a next step, the determined results should be transferred to other distribution types, e.g. lognormal, uniform or triangular distribution.

4 Methods for Analysing Uncertainties in LCA

There are several methods for calculating uncertainties in LCI data and LCA results. Three main methods, the Monte Carlo simulation, the pedigree approach (an example for data quality indicators), and the sensitivity analysis (commonly used in LCA practice) are described in the following section.

When performing a Monte Carlo simulation, the characterizing probability density functions for each in- and output of an LCI/LCA model are used. The in- and output values are taken as random numbers, and then the LCI is calculated. This procedure is repeated many times, the so-called iterations. The LCI result is stored for each iteration. Hence, the probability density function of the results can be estimated and the confidence interval can be calculated. There are two disadvantages to this approach, though: First, the result's distribution is estimated only, based on a random number generator, second, this method requires a lot computation time (Feck 2007; Goedkoop 2016).

When estimating the data uncertainty with the pedigree matrix, five categories and scores are used. The pedigree matrix values determine (part of) the flow's variance. This method is not used for uncertainty estimations within the LCI- or LCIA results, but it can be used supplemental to the Monte Carlo simulation method (Weidema et al. 2013).

The sensitivity analysis is generally used to investigate model and scenario uncertainties. Individual parameters or decisions within an LCA/LCI model can be modified, the result is calculated and the effect on the outcome is observed. This method is in particular used for the estimation of different modeling assumptions while the other methods mentioned are used for the estimation of data variability. (Norm DIN EN ISO 14040:2009–2011; Norm DIN EN ISO 14044:2006–2010).

5 Convolution as an Approach

This approach is based on the fact, that the probability density function of the sum of two, independent, continuous random variables is the convolution of the probability density functions of each variable. Equation (4) shows the calculation, where t is the independent variable of the resulting function, u is the integration variable, f_X is the probability densitiy function of the first in-/output and f_Y is the probability density function of the second in-/output. The convolution of more than two elements can be computed successively because of the fact that the convolution is associative (Holzmann et al. 2011; Weyerhäuser 2005).

$$f_{X+Y}(t) = (f_X * f_Y)(t) = \int\limits_{-\infty}^{+\infty} f_X(t-u) \cdot f_Y(u)du \tag{4}$$

The following assumption is made in order to transfer the convolution theory to an LCA model: The LCI data are independent, continuous and normally distributed. The advantage of the normal distribution is, that the convolution result of two normal distributions is also a normal distribution (Hübner 2009). Nevertheless, the convolution and the combination of different distribution types should be included in case of further methodological development. Furthermore, the LCA model is a linear system, which is described with Eq. (5) (IAF: specific impact assessment factor of an elementary or product flow, WI: impact category, X is one elementary/product flow, an Y is the second one). Thus, LCA-models can also be non-linear, so-called complex, systems. The description of a complex system as a linear system belongs to model uncertainty, which is not yet analysed within this research.

$$WI = IAF_X \cdot X + IAF_Y \cdot Y \tag{5}$$

The determination of the impact category's probability density function, a combination of Eqs. (4) and (5), is shown in Eq. (6).

$$f_{WI}(t) = \int\limits_{-\infty}^{\infty} \frac{1}{IAF_X} \cdot f_X\left(\frac{u}{IAF_X}\right) \cdot \frac{1}{IAF_Y} \cdot f_Y\left(\frac{t-u}{IAF_y}\right)du \tag{6}$$

The illustrated considerations about the convolution are transferred to a virtual, small-LCA system. This model is shown in Fig. 2.

Fig. 2 Illustration of a virtual, small LCA-model

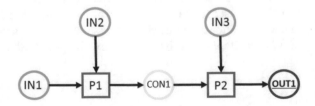

The impact category result of the production of one piece of reference product equals the specific impact assessment factor (IAF) of this product. CON 1 is the reference product of process P 1. So, the IAF's probability function of CON 1 is determined by the convolution of the probability density functions of elementary flow IN 1 and IN 2 and the IAF's probability function of OUT 1 is calculated by the convolution of the probability density functions of elementary flow IN 3 and product flow CON 1. The calculation rules are shown in Eqs. (7) and (8).

$$f_{WI}(t) = \frac{1}{IAF_{IN3}} \cdot \frac{1}{IAF_{CON1}} \cdot \int_{\mathbb{R}} f_{IN3}\left(\frac{u}{IAF_{IN3}}\right) \cdot f_{CON1}\left(\frac{t-u}{IAF_{CON1}}\right) du \qquad (7)$$

The IAF_{CON1} is emphasized in Eq. (7) because it is the result of the convolution of the elementary flow's distributions of process P1 [see also Eq. (8)].

$$f_{IAF_{CON1}}(t) = \frac{1}{IAF_{IN1}} \cdot \frac{1}{IAF_{IN2}} \cdot \int_{\mathbb{R}} f_{IN1}\left(\frac{u}{IAF_{IN1}}\right) \cdot f_{IN2}\left(\frac{t-u}{IAF_{IN2}}\right) du \qquad (8)$$

Hence, we conclude, that the elementary flow's parameter uncertainty propagate via the product flow's specific impact assessment factor and have an effect on the LCA-model result.

6 Outlook

The presented work shows a first approach and requires more research. The following points illustrate further work:

The assumption of uncertainty propagation through the specific elementary flow's IAF should be proved based on adequate models. Furthermore, it should be worked out, to what extent a large dispersion of a product flow's IAF has influence of the impact category's one. In addition to that, the minimization of this dispersion should also be analysed.

The method mentioned should be transferred to probability density function types other than the normal distribution, because the parameter uncertainty of elementary and product flows can also have other distribution types, like the uniform, triangular or lognormal distribution. Furthermore, it should be usable, if two or more distribution types are mixed.

The results of the new method should be compared to the results of the Monte Carlo simulation, since the latter is the prevailing method, which is included partially in LCA software.

The computational structure of an LCA model consists of matrices. Because of this, the method should be translated to the model of the computational structure of LCA, which was described by Heijungs and Suh (2002).

Beyond investigating the parameter uncertainty, a method should be found to include scenario as well as model uncertainty.

Currently, there is no general method for analysing uncertainties in LCA case studies. Either uncertainties are not mentioned at all or different methods are used. In a future research, a method should be developed, which concludes the advantages of all prevailing methods (Lloyd and Ries 2007). In the best case, this method should be admitted as standard uncertainty analysing method. All LCA authors should interpret its results correctly and communicate them in an understandable way, so they could be the basis for important decisions, e.g. within the economy.

References

Bourgault, G.: How to Interpret the Uncertainty Fields in Ecoinvent? Ecoinvent Association. Zürich. Available online at http://www.ecoinvent.org/support/faqs/methodology-of-ecoinvent-3/how-to-interpret-the-uncertainty-fields-in-ecoinvent.html. Last updated 02 Dec 2016, last checked 02 Dec 2016 (2016)

Bourgault, G.: Reasons for the Choice of Probability Density Functions of LCI-Data. Zürich. Available online at http://www.ecoinvent.org/support/ecoinvent-forum/topic.html?&tid=270. Last updated 26 Feb 2016, last checked 26 Feb 2016 (2017)

Feck, N.: Monte-Carlo-Simulation bei der Lebenszyklusanalyse eines Hot-Dry-Rock-Heizkraftwerkes. Dissertation, Ruhr-Universität Bochum, Bochum. Fakultät für Maschinenbau (2007)

Goedkoop, M. et al.: SimaPro Tutorial. PRé (5.3) (2016)

Hauschild, M., Huijbregts, M.A.J. (Hg.): Life Cycle Impact Assessment. Springer (LCA compendium, the complete world of life cycle assessment), Dordrecht. Available online at http://search.ebscohost.com/login.aspx?direct=true&scope=site&db=nlebk&AN=970501 (2015)

Hedderich, J., Sachs, L.: Angewandte Statistik. Methodensammlung mit R. 15., überarb. u. erweiterte Aufl. Springer Spektrum, Berlin (2015)

Heijungs, R., Suh, S.: The Computational Structure of Life Cycle Assessment, p. 11. Kluwer Academic Publishers (Eco-efficiency in industry and science, Boston (2002)

Holzmann, G., Meyer, H., Schumpich, G., Eller, C., Dreyer, H.-J.: Technische Mechanik. Statik. 12., verb. Aufl., unveränd. Nachdr. Vieweg + Teubner (Studium,/Günther Holzmann; Heinz Meyer; Georg Schumpich), Wiesbaden

Hübner, G.: Stochastik. Eine anwendungsorientierte Einführung für Informatiker, Ingenieure und Mathematiker. 5., verb. Aufl. Vieweg + Teubner (Studium), Wiesbaden (2009)

Huijbregts, M., Heijungs, R. et al.: Complexity and Integrated Resource Management: Uncertainty in LCA. Int. J. Life Cycle Assess. (5), S. 341–342 (2004)

Koch, A.L.: The logarithm in biology. Mechanismus generating the log-normal distribution exactly. J. Theoret. Biol. 12, 276–290 (1966)

Limpert, E., Ohmayer, G., Stahel, W.A.: Eine Grundlage der Datenanalyse: Addiert oder multipliziert die Natur? BioScience 51(5), S. 341–352. Available online at http://www.gil-net.de/Publikationen/24_191.pdf. Last checked 02 Dec 2016 (2001a)

Limpert, E., Stahel, W.A., Abbt, M.: Log-normal distributions across the science: keys and clues. Bioscience 51(5), 341–352 (2001b)

Lloyd, S.M., Ries, R.: Characterizing, Propagating, and analyzing uncertainty in life-cycle assessment. A survey of quantitative approaches. J. Ind. Ecol. 11(1), S. 161–179 (2007)

Menzel, C.: Unsicherheiten bei Ökobilanzen mit Umberto/ecoinvent. unveröffentlichtes Manuskript. Hochschule Niederrhein. Mönchengladbach (2016)

Muller, S., Lesage, P., Ciroth, A., Mutel C., Weidema, B.P., Samson, R.: The application of the pedigree approach to the distributions forseen in ecoinvent v3. Int. J. Life Cycle Assess. 21, 1327–1337 (2016)

Mutel, C.: Why does the ecoinvent database love the lognormal distribution? Zürich. Available online at https://chris.mutel.org/ecoinvent-lognormal.html. Last updated 12 June 2016, last checked 12 June 2016 (2013)

Opitz, A.: Vergleichende Sachbilanz von Tofu und Schweinefleisch von der Erzeugung bis zur Zubereitung. Bachelorarbeit, Hochschule Niederrhein, Mönchengladbach. Fachbereich Oecotrophologie (2016)

Publication Office of the European Union (Hg.): General guide for Life Cycle Assessment. Detailed guidance. In: ILCD Handbook—International Reference Life Cycle Data System. European Commission-Joint Research Centre—Institute for Environment and Sustainability. Luxemburg (2010)

Qin, Y., Suh, S.: What distribution function do life cycle inventories follow? Int. J. Life Cycle Assess. (2016)

Suh, S., Leighton, M., Tomar, S.: Interoperability between ecoinvent ver. 3 ans US LCI database: a case study. In: Int. J. Life Cycle Assess. 21, S. 1290–1298. Available online at https://link.springer.com/article/10.1007/s11367–013-0592-2/fulltext.html. Last checked 12 Jan 2017 (2016)

Norm DIN EN ISO 14044:2006–10: Umweltmanagement – Ökobilanz – Anforderungen und Anleitungen (2006)

Norm DIN EN ISO 14040:2009–11: Umweltmanagement – Ökobilanz – Grundsätze und Rahmenbedingungen (2009)

Weidema, B.P., Bauer, C., Hischier R., Mutel C., Nemecek, T., Reinhard, J. et al.: The ecoinvent Database: Overview and Methodology. Data quality guideline for the ecoinvent database version 3. Hg. v. ecoinvent centre. Swiss Centre for Life Cycle Inventories. St. Gallen. Available online at www.ecoinvent.org. Last checked 18 Oct 2015 (2013)

Weyerhäuser, K.: Faltung und Korrelation kontinuierlicher Signale. Universität Koblenz, Koblenz. Institut für integrierte Naturwissenschaften Abteilung Physik, Seminararbeit (2005)

LCA of Energy and Material Demands in Professional Data Centers: Case Study of a Server

F. Peñaherrera, J. Hobohm and K. Szczepaniak

Abstract Professional Data Centers (PDC) quantity and their related energy consumption is in a continuous growing trend, with corresponding increments in the material demand. While the amount of energy during operation phase is normally measured, the amount of embedded energy and the total Cumulated Energy Demand (CED) is often overlooked. To estimate the total amount of energy and material demand, LCA is used to evaluate the CED and CMD (Cumulated Material Demand) of devices used in PDC. A server Unit typical of these facilities is analyzed using data from disassembly and elementary analysis to construct a model for LCA to estimate the CED and CMD of the device. Results indicate that most of the energy is consumed during the operation phase, which is dependent on the operating lifetime of the device. Uncertainties are present due to the quality of the databases used for the model.

Keywords LCA · Data Center · Energy Efficiency · Material Demand

1 Introduction

More than 12 TWh/a are consumed by around 53,000 PDC in Germany (Fichter and Hintemann 2014). This energy consumption is expected to grow to 14 TWh/a by 2020 (Hintemann 2016). Information and Communication Technology (ICT) products have a short lifetime between 3 to 5 years (Garnier 2012), indicating that the energy consumption outside of the service phase is also of relevance. The number of servers grew by 28% for the period 2010–2014, with the trend to build bigger data centers (Hintemann and Clausen 2014). This is accompanied by the increasing material intensity in ICT, so the material demand for these applications rises in parallel

F. Peñaherrera (✉)
Carl von Ossietzky University of Oldenburg, Oldenburg, Germany
e-mail: fernando.andres.penaherrera.vaca@uni-oldenburg.de

J. Hobohm · K. Szczepaniak
Technische Universität Hamburg, Hamburg, Germany

© Springer Nature Switzerland AG 2019
L. Schebek et al. (eds.), *Progress in Life Cycle Assessment*, Sustainable
Production, Life Cycle Engineering and Management,
https://doi.org/10.1007/978-3-319-92237-9_9

to the energy consumption. The TEMPRO Project (Total Energy Management for Professional Data Centers), financed by the Federal Ministry for Economic Affairs and Energy (BMWi) in the framework of the 6th Energy Research Program of the Federal Government, has as one of its objectives the creation of an assessment basis for the holistic analysis of energy and resource efficiency of PDC.

The objective of this analysis is to assess the amount of energy consumed by a unit of a PDC equipment, a server, in its various stages. Emphasis is placed on analyzing the dependencies of the results on embodied energy in combination with the data on the energy consumption for the operation phase of a ICT device with the results on the evaluation of the quantities of materials required for the production and operation phases. The concept of embodied energy attached to the steps of raw parts production, manufacturing, transporting, decommissioning and recycling are often excluded when analyzing energy consumption or improvements on energy efficiency, which focuses mostly on the use phase, and overlooks the importance of the critical material content of ICT applications. Studies into the energy use of ICT devices during its life cycle prove that as the equipment becomes more operationally efficient, the embodied stage will play a larger role in the full life cycle (Whitehead et al. 2015).

The goal of this analysis is to assess the energy and material consumption of a server during its lifetime. For this purpose, the concepts of CMD and CED are applied (Giegrich et al. 2012). Existing gaps in the data availability are considered and attempted to close in some areas, through technology comparison and Pedigree Analysis.

2 Methodology

2.1 PDC Components

The structure of a PDC differs from each other. Common classifications of PDC characterize them by size (surface area, m^2), by ICT installed capacity (kW), or by total installed capacity (including all facilities, kW) (Fichter et al. 2010). This information excludes data on the number and type of devices, or on the various aspects of the PDC configuration, such as redundancy, cooling technologies and type, safety, or energy sources. Additional information relevant to the estimation of the CED and CMD, such as individual energy consumption, operating lifetime, or energy conversion efficiency of the power supply infrastructure, are normally unavailable. To analyze the energy indicators of a PDC, an individual analysis of the various components is to be developed to obtain single indicators that can be aggregated and extrapolated for an estimation of the total indicators.

The different elements can be categorized according to their application and role in a PDC: ICT, Electrical Power Supply, Climatization, and Infrastructure (Fig. 1).

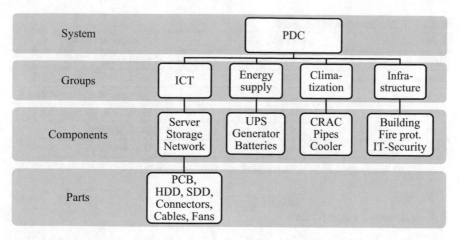

Fig. 1 Components of a Professional Data Center. *Source* Modified from Szczepaniak and Hobohm (2017)

The first portion of the analysis focuses on analyzing ICT devices, due to their short operating lifetime and their energy and material intensity. A server for PDC applications is disassembled to analyze its parts composition and material content. With the parts and elements inventory, a model for LCA is constructed to analyze the CED and the CMD. Due to the quality of the data obtained, the uncertainty of the results is assessed using Pedigree Analysis and Monte Carlo simulations.

2.2 Goal and Scope Definition

The objective of the LCA of the server unit is to analyze the CED and CMD during its lifecycle. Different stages are considered: parts production, assembly, and operation. The scope is limited to the end of the operation time, since no recycling strategies or End-of-Life procedures are specified. For the operation of the unit, energy inputs for the cooling and for the service are considered.

The functional unit is the life-time of the device, which is set to 5 years. To calculate the indicators, data on unit composition, parts and weight is required, as well as technical data on power consumption and operating lifetime. A model is then build in openLCA 1.6, which allows also assessing the uncertainties via Pedigree Analysis and Monte Carlo simulations. Data is then analyzed using MATLAB R 2017a.

Table 1 Bill of materials of the server

Denomination	Weight (%)	Principal materials
Housing	50.0	Steel, Al
Storage	21.8	Steel
Power supply	13.2	Ceramic, steel, Cu, plastics
PCB	10.3	Plastic, Cu, Al, ceramics
Fans	4.3	Steel, plastic, Cu
Cable	0.4	Cu, plastic

Source Elaborated with data from Szczepaniak and Hobohm (2017)

2.3 Inventory Analysis

A server model hp ProLiant DL 380 is selected for analysis. The technical characteristics of the unit indicate a maximum rated input power of 1170 W, and a heat
generation 3990 BTU/h (hp 2006). A disassembly of the unit is performed to obtain
a component list with the relative weight of the parts (Table 1). These different parts
are then analyzed to obtain a detail of the material and elemental composition of the
unit (Szczepaniak and Hobohm 2017) (Fig. 2).

The system boundaries for the analysis include the energy used for transporting
of the pieces, manufacturing, operation and cooling of the device. End-of-Life is not
considered (Fig. 3).

The characteristic operation of a PDC is one of continuous operation mode with
fluctuations on the power consumed by the ICT devices, and with cooling dependent
on the outdoor conditions (which may allow the use of free cooling). ICT devices
operate in a state between idle mode and maximum power consumption depending
on the load. The average load for a device can be estimated as 70% of their maximum
load (Rasmussen 2013). The number of operation hours during the year in a PDC are
relative high, with an unavailability lower than 12 h/a (BITKOM 2013). This allows
estimating the yearly energy consumption of the device, and use it as input for the
energy consumption.

The database (DB) used in the software for calculations is the ecoinvent 3.3
(2017). Since the specific parts materials are absent in the DB, these are matched
with existing processes. A Pedigree Analysis (Ciroth et al. 2016) is incorporated to
assess the quality of the DB, assigning a matching geometric uncertainty (σ_g) of a
normal logarithmic distribution to the product model (Fig. 4; Table 2).

3 Results and Discussion

The results of the CED indicate a prevalence of the energy required for operation and
cooling of the device (50.7 and 39.6% respectively, Fig. 5). The primary energy for

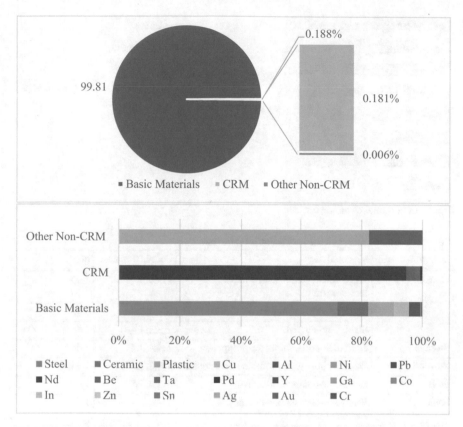

Fig. 2 Material composition. *Source* Elaborated with data from Szczepaniak and Hobohm (2017)

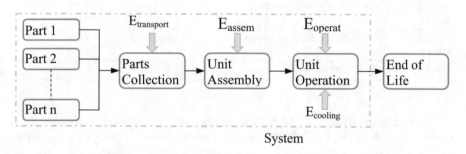

Fig. 3 Process diagram for LCA of a server

operation is 3.1 times the input electric energy to the device, representing conversion and transformation losses of energy. This without consideration of the energy losses in the UPS (Uninterruptible Power Supply) and transformers belonging to the PDC infrastructure, which were not part of the analysis.

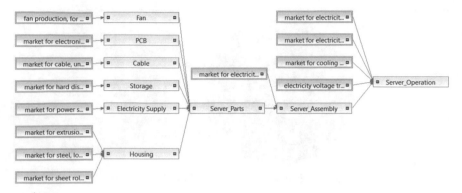

Fig. 4 LCA model of the server

Table 2 Correspondence to the ecoinveint 3.3 database

Denomination	Modeled as	Pedigree matrix	σ_g
Housing	Extrusion of plastic sheets and thermoforming	1;3;3;3;4	1.45
	Sheet rolling, aluminum	1;3;2;3;4	1.45
	Steel, low-alloyed	1;3;2;3;4	1.45
Storage	Hard disk drive, for laptop computer	1;4;4;3;4	1.46
Power Supply	Power supply unit, for desktop computer	1;3;3;4;5	1.68
PCB	Electronic component, active	1;3;4;3;5	1.69
Fans	Fan, for power supply unit, desktop computer	1;3;2;3;4	1.45
Cable	Cable, unspecified	1;3;2;3;2	1.10
Assembly energy	Electricity, high voltage	4;3;3;3;3	1.15
Operation energy	Electricity, medium voltage	4;3;1;4;3	1.44
Cooling energy	Cooling energy	4;3;1;4;4	1.58

The energy involved in the fabrication and assembly of other components occupy a lower fraction of the CED, 9.7%. This relationship between energy for operation and for fabrication is dependent on the service lifetime of the device, and reducing it increases the significance of the CED outside the operation phase.

The Monte Carlo Analysis provides a visualization of the error propagation resulting of the different uncertainties of the input data, presenting a logarithmic normal distribution (Fig. 6). The performed laboratory analysis to characterize the material composition of the device improves the data quality, providing high certainty in the input values of the weight for parts composition, and therefore reducing the results uncertainty. There is little data on the production of components and End-of-Life processes. Data from the ecoinvent DB must be compared with similar technologies

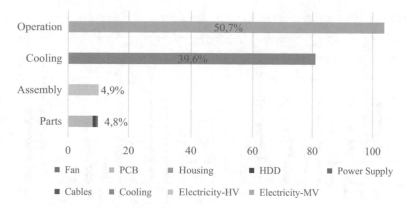

Fig. 5 CED fort the analyzed server, MWh

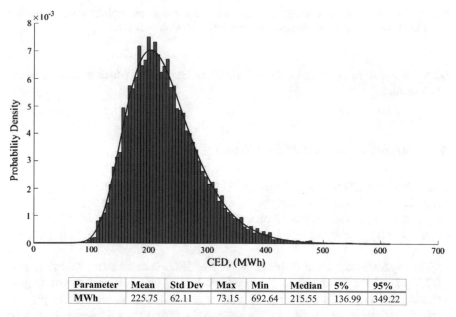

Parameter	Mean	Std Dev	Max	Min	Median	5%	95%
MWh	225.75	62.11	73.15	692.64	215.55	136.99	349.22

Fig. 6 Monte Carlo analysis for the CED of the server. n = 10 000

(laptops of laptops, HDD of PCs), increasing uncertainties, and thus increasing the overall error in the results. The resulting standard deviation is 27.51% of the mean value.

The analysis of the material requirements for the server is performed through an LCA focusing on the material consumption to characterize the CMD. Due to the different nature of the components in the database, the results of certain materials differ from those obtained in laboratory analysis (Fig. 7). This is another indicator of the requirement to update the datasets. There are materials resulting from the

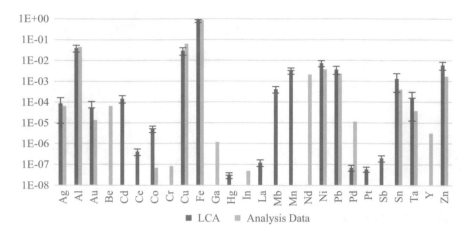

Fig. 7 CMD of the server obtained from LCA and from composition analysis. Values in fraction of the total. *Source* Analysis data from Szczepaniak and Hobohm (2017)

laboratory analysis, such as Gallium, Indium, or Beryllium, which are absent in the LCA results.

4 Conclusions and Further Work

The values of CED resulting from the LCA show the importance of the operation phase in the total energy consumption, this in link with the lifetime of the device. A major part of the CED outside the operation phase is the fabrication of PCB (printed circuit boards), with is also of relevance in the End-of-Life stages due to its material content.

The logarithmic normal distribution of the CED presents a standard deviation of 27.51% of the mean value. This is the result of using databases that require updating and improved matching. A more recent database is required to provide higher certainty in the results. In this case study, different technologies had to be matched with the provided data, so that the model is built to represent the device. Constant sources of uncertainties were the dissimilarity of the technologies, the representativeness of the datasets used, and the temporal correlation of the data sources.

The results on the material content analyzed can be further used to evaluate material efficiencies in the processes of manufacturing the devices, since the difference in the materials present in the device and the material quantities resulting from the CMD analysis indicate material losses during the lifetime. This in conjunction with an analysis for reuse and recycling of the materials present at the End-of-Life of the devices can increase the material and energy efficiency.

Only individual components were analyzed. Future work will continue the analysis of key single components of a PDC and develop indicators for their aggregation for a complete evaluation of such facilities. The extrapolation of results for complete PDC can be evaluated as soon as sufficient indicators are available. Scenarios for the End-of-Life of the various devices that contain CRM will be analyzed when more data regarding the End-of-Life stages is available, so that estimations of energy savings and material gains are calculated to develop appropriate recycling strategies. The optimum replacement of older devices with considerations of the CED is to be investigated as the replacement is usually made because of the expiration of the warranty on the maintenance of the equipment.

References

BITKOM: Betriebssicheres Rechenzentrum. BITKOM. Berlin, Germany (2013). Available online at https://www.bitkom.org/noindex/Publikationen/2013/Leitfaden/Betriebssicheres-Rechenzentrum/LF-Betriebssicheres-Rechenzentrum.zip, checked on 1/11/2018

Ciroth, A., Muller. S., Weidema, B., Lesage, P.: Empirically based uncertainty factors for the pedigree matrix in ecoinvent. In Int. J. Life Cycle Assess. **21**(9), 1338–1348, checked on 1/11/2018 (2016)

Fichter, K., Hintemann, R.: The quantities of materials present in the equipment of data centers. J. Indus. Ecol. (2014). Available online at http://onlinelibrary.wiley.com/doi/10.1111/jiec.12155/epdf, checked on 1/11/2018

Fichter, K., Hintemann, R., Stobbe, L.: Materialbestand der Rechenzentren in Deutschland -. Eine Bestandsaufnahme zur Ermittlung von Ressourcen- und Energieeinsatz. Dessau-Roßlau, Germany (2010). Available online at https://www.umweltbundesamt.de/sites/default/files/medien/461/publikationen/4037.pdf, checked on 1/11/2018

Garnier, C.: Data Centre Life Cycle Assessment Guidelines. The Green Greed. Oregon, USA (2012). Available online at https://www.thegreengrid.org/en/resources/library-and-tools/236-Data-Center-Life-Cycle-Assessment-Guidelines, checked on 1/1/2018

Giegrich, J., Liebich, A., Lauwigi, C., Reinhardt J.: Indikatoren/Kennzahlen für den Rohstoffverbrauch im Rahmen der Nachhaltigkeitsdiskussion. Edited by UBA. UMWELTBUNDESAMT. Dessau-Roßlau, Germany (2012). Available online at http://www.uba.de/uba-info-medien/4237.html, checked on 1/11/2018

Hintemann, R.: Energy consumption of data centers continues to increase. 2015 update. Borderstep Institute. Berlin, Germany (3) (2016). Available online at https://www.borderstep.de/wp-content/uploads/2015/01/Borderstep_Energy_Consumption_2015_Data_Centers_16_12_2015.pdf, checked on 1/11/2018

Hintemann, R., Clausen, J.: Rechenzentren in Deutschland. Eine Studie zur Darstellung der wirtschaftlichen Bedeutung und der Wettbewerbssituation. Borderstep Institut. Berlin, Germany (2014). Available online at https://www.bitkom.org/noindex/Publikationen/2014/Studien/Studie-zu-Rechenzentren-in-Deutschland-Wirtschaftliche-Bedeutung-und-Wettbewerbssituation/Borderstep-Institut-Studie-Rechenzentren-in-Deutschland-05-05-20141.pdf, checked on 1/11/2018

hp: HP ProLiant DL380 Generation 5 (G5). QuickSpecs (2006). Available online at https://h20195.www2.hpe.com/v2/getpdf.aspx/c04282492.pdf?ver=4, checked on 1/11/2018

Rasmussen, N.: Calculating Space and power Density Requirements for Data Centers. Schneider Electric (2013). Available online at http://www.apc.com/salestools/NRAN-8FL6LW/NRAN-8FL6LW_R0_EN.pdf, checked on 1/10/2018

Szczepaniak, K., Hobohm, J.: Quantifizierung kritischer Rohstoffe in Rechenzentren als ressourcenrelevanter Bezug zur Grauen Energie. (Unveröffentlicht). Master Thesis. Tecnische Universität Hamburg, Hamburg, Germany (2017)

Whitehead, B., Andrews, D., Shah, A.: The life cycle assessment of a UK data centre. Int. J. Life Cycle Assess. **20**(3), 332–349 (2015). https://doi.org/10.1007/s11367-014-0838-7

Identification of Potentials for Improvement in Paint Production Process Through Material Flow Cost Accounting—A Step Towards Sustainability

Esther Peschel, André Paschetag, Mandy Wesche, Helmut Nieder and Stephan Scholl

Abstract During the development of sustainable processes ecological as well as economic aspects play an increasingly important role alongside the social factors. These can be taken into account by using material flow cost accounting (MFCA) according to ISO 14051 which captures the environmental and economic impacts of material and energy inputs. In the BMWi collaborative research project µKontE the transfer of a batchwise to a continuous production of binder emulsion was investigated. A MFCA was set up as a case study for both manufacturing strategies. The influence and benefits of different levels of detail in the definition of the quantity centres were examined.

Keywords Material flow cost accounting · Energy and resource efficiency Batch to conti · Process development · 3-Level-Model

1 Introduction

In process development, sustainability has become the focus of industry and science with the goal of decoupling economic growth from the increasing consumption of resources (Hirzel et al. 2013). In 1987 the United Nations defined the concept of "sustainable development" as part of the Brundtland report. According to this, it is a

E. Peschel (✉) · A. Paschetag · M. Wesche · S. Scholl
Technische Universität Braunschweig, Institut Für Chemische und Thermische Verfahrenstechnik, Brunswick, Germany
e-mail: e.peschel@tu-braunschweig.de

H. Nieder
AURO Pflanzenchemie AG, Brunswick, Germany

© Springer Nature Switzerland AG 2019
L. Schebek et al. (eds.), *Progress in Life Cycle Assessment*, Sustainable Production, Life Cycle Engineering and Management,
https://doi.org/10.1007/978-3-319-92237-9_10

"development that meets the needs of the present generation, without jeopardising the ability of future generations to meet their own needs" (BMUB 2018). Subsequently three dimensions of sustainability, ecology, economy, and social issues, were derived from the concept of "sustainable development" and transferred to the so-called "three-pillar model" (von Hauff and Kleine 2014). Given a technology-wise feasible and operable process, an incorporation of economic, ecological and social aspects has become mandatory in today's process development.

Material flow cost accounting (MFCA) is one possible approach which can be used to evaluate the resource efficiency and take the ecological and economic aspects into account. The MFCA was developed in the late 1990s and is already widely used in different industries (Schrack 2016; METI 2010). In 2011 it was officially released as an international standard to serve as a management tool in organizations to consider "potential environmental and financial consequences of their material and energy use practices" (ISO 14051:2011). Compared to conventional cost accounting, MFCA allocates the expenses both to the products and to the material and energy losses ("non-products"). In its overall objective, MFCA aims at the quantification and visualization of material losses, thereby contributing to the motivation of the organization to reduce material and energy consumption (Viere et al. 2009, ISO 14051:2011). By reducing the use of materials and energy, on the one hand, the costs and, on the other hand, the impact on the environment should be reduced. The eco-logical aspect is thus indirectly taken into account in the MFCA. The material losses are only recorded in physical and monetary units and can pollute the environment as waste, sewage or emissions. Unlike the Life Cycle Assessment (LCA), MFCA does not provide an impact assessment, so the potential direct environmental impacts are not investigated. MFCA is primarily used in the evaluation of productions of general cargo at the product and plant level (METI 2010). In the material-converting industry as well, the application of MFCA can be used to identify potentials for improvement with regard to the use of energy and raw materials.

In the German Federal Ministry for Economic Affairs and Energy (BMWi) research project μKontE,[1] a production process of a binding agent emulsion was transferred from a batch to a continuous mode of operation. First of all, the concep-tion of the production plant and then the establishment in production took place. The initial batch process as well as the newly developed continuous process were evalu-ated and compared among others through MFCA. To investigate the applicability of MFCA during process development, the influence and benefits of different levels of detail in regards to the process modelling were examined.

---

[1] Increasing the energy efficiency of production processes by transfer from batchwise to continuous production using mini- and micro-process components—μKontE 03 ET1093A.

2 Description of the Process

AURO Pflanzenchemie AG provided the investigated binding agent emulsion production process. For this case study process data, e.g. energy consumption, type and amount of process material and reactants, and information relating to the different cost types from the company were used. The batch production was carried out in an agitator vessel with heating jacket in which all process steps were performed consecutively. At the start, all reactants, consisting of partially organic as well as aqueous chemicals, are fed to the reaction vessel and heated up. Afterwards, several emulsification and shearing steps are applied until the required droplet size is achieved. Finally, the binding agent emulsion is drained from the stirred tank and the equipment is cleaned. In continuous production, the respective preparation of the organic and the aqueous reactants takes place simultaneously outside of the actual reaction vessel. These two phases are then combined and pre-emulsified in a passive mixer (premixer). Subsequently, this pre-emulsion is emulsified in an active mixer (rotor-stator mixer) in order to achieve the required product properties.

The transfer of this process from batch to continuous production involved several constraints with respect to the manufacturing scheme. The advantages of batch production concerning the production flexibility due to fluctuating market demands had to be maintained at the same time ensuring a constant and high product quality through online monitoring of process parameters in the continuous procedure. Furthermore, the high energy consumption of the batch production due to consecutive heating and cooling steps in a single vessel were targeted for heat integration in the continuous production process.

It was shown, that the specific energy demand in continuous production has been reduced by 50% compared to the batch process (Wengerter et al. 2017). While batch production required a high level of cleaning effort, the continuous operation of the system could be significantly reduced through the implementation of microstructured devices and thus lower plant hold-up and significantly increased cleaning intervals. Furthermore, personnel expenditures were considerably reduced by a largely automated plant control for the continuous process.

The MFCA was conducted in accordance with ISO 14051: 2011–2012. Its application was intended to increase energy and resource efficiency and to support plant and equipment design in the case study. The increase in energy and resource efficiency can be achieved on the one hand by an appropriate process concept, such as the transfer from batch to continuous production, and on the other hand by the choice of equipment. The MFCA can be applied to evaluate both aspects depending on the definition of the quantity centres. In the context of this article, MFCA focuses on the choice of equipment and its implementation to increase energy and resource efficiency and improve plant and equipment design. The definition of the quantity centres was based on the 3-level-model according to (Wesche et al. 2015). In the first level , the model maps the individual unit operations (UO) of the production

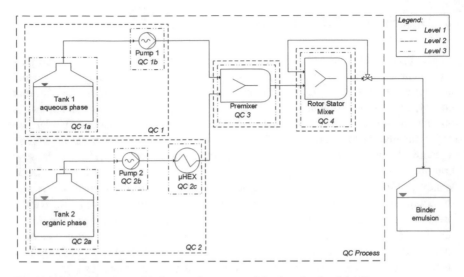

Fig. 1 Process flow chart with the quantity centres of the three levels of detail

process, which are connected to create the full process topology in the second level. The available infrastructure (operating networks, disposal routes, etc.) are recorded in the third level. The plant for continuous binder production with the individual observation levels for MFCA is shown in Fig. 1 (ISO 14051:2011).

According to ISO 14051, the costs of the materials, the energy, the system and the waste management have to be taken into account. The material costs were allocated based on the mass balance for the individual quantity centres. The energy costs were allocated to the quantity centres according to the respective energy consumption. In this case, the system costs accounted for the monetary expenditures for intra-logistics, maintenance and the machine hourly rate as well as personnel costs. The first three cost items were allocated according to the number of quantity centres, while personnel costs were allocated according to the duration of work per quantity centres. The material losses were accounted for 100% of the waste management costs. These cost elements and their allocation have been taken into account for the different levels of detail with regard to the quantity centres.

The examined first level of detail included the production process as a single quantity centre, so that the input and output flows as well as costs were recorded in total over the full production plant. The second level included the structure of the plant by breaking down the model into the individual process steps. In the third level of detail, the individual components of the plant equipment of the respective UO were each assigned to individual quantity centre in order to enable the analysis of the plant and apparatus design in regards to potential improvements. Above all, the investigation aimed at a possible increase in energy and resource efficiency with regard to instrumental implementation.

3 Results and Discussion

The investigation on the first level of detail showed that material costs accounted for more than 90% of the overall cost. Thus, one option to reduce the cost of materials is the use of cheaper, alternative reactants. Although cheaper reactants may have a positive effect on manufacturing costs, they may also have a negative effect on product quality due to higher impurity levels. System costs accounted for the second highest share of costs at 9.31%, while energy costs are comparatively low at 0.58%. However, both system costs and energy costs are in absolute terms a big expense for the company, so they should be investigated further.

For the second level, four quantity centres were considered according to the process steps, as shown in Fig. 2. The material costs are mainly caused by the organic phase, while the costs for the aqueous phase only make up a small proportion of the costs. Therefore, deviations from the aspired product composition of the organic and aqueous phase have significant effects on the material costs. One possible approach to reducing material costs is the use of alternative, cheaper reactants. However, according to the company philosophy, these must be natural substances, so that the choice of alternatives is very limited. In addition, the revision of a recipe in an existing production process is economically not reasonable for a company. Furthermore, the second level of detail revealed the highest energy and system costs in the preparation of the organic phase compared to the other quantity centres.

Lastly, the highest level of detail is displayed in Fig. 3. It shows that the highest energy costs are caused by the pump of the organic phase. In comparison to the aqueous phase, it has a comparatively high viscosity leading to a higher energy demand for pumping. If necessary, these energy costs can be reduced by a more efficient pump, whereas this is associated with investment. Therefore, it is important to examine

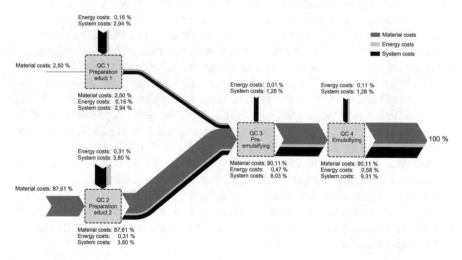

Fig. 2 Cost flow of the second level

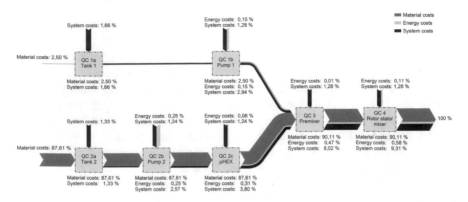

Fig. 3 Cost flow of the third level

the interaction between the cost of acquisition and the potential savings in energy costs. The third level also shows, contrary to expectations from the second level, that the highest system costs are caused by the quantity centre 1a, which represents the preparation of the aqueous phase. Personnel costs are allocated to the system costs according to the number of quantity centres based on the duration of work per quantity centre, the increased costs at this point are due to the increased personnel expenses. These monetary expenses can be reduced for example by automated dosing of the aqueous reactants. However, this is again associated with increased costs, which are caused on the one hand by the acquisition but also by the operation and maintenance of the additional equipment.

The investigations show that the effort for the necessary data collection increases significantly with the increase in the level of detail. The process data for each level of detail were collected directly at the system. For example, the energy consumption for each equipment was recorded by means of appropriate measuring points. However, this depends on the particular system, the measuring equipment and the procedures within the company itself. The third level is of great interest from an engineering point of view, as both the materials used and the design can have a direct influence on the required energy input. Thus, in combination with information about the individual UO process, engineering potentials can be identified, which can serve as a basis for technical improvements. In addition, findings from the detailed presentation of the system costs, in particular with regard to the space costs as well as maintenance and servicing depending on the technology used, are relevant in this context.

However, such an extensive analysis of the process is not needed in many cases. For example, in the development phase, data about the individual equipment is an advantage, while less detailed information about the production process are required for the management to support decision-making. Therefore, the level of detail should be tailored to the information needs of the target group based on the workload prior to the implementation of the MFCA.

The MFCA takes the ecological aspects into account indirectly. The material and energy inputs as well as the material losses can be assigned to their causes. However, MFCA does not identify the direct impact on the environment but merely identifies possible starting points for improvements in resource use. For a closer examination of the ecological aspects, a linkage with the LCA should be made. For a full view of sustainability, in addition to MFCA and LCA, social aspects must be considered. These are reflected in the production, for example, by the commitment and the tasks of the employees. This can be investigated, for example, by means of social life cycle assessment (SLCA) (Klöpffer and Grahl 2009).

4 Summary

As part of the μ KontE project, a process for the production of a binding agent emulsion has been transferred from batch to continuous operation. Through the implementation of a development accompanying MFCA, the process transfer was supported and evaluated. Furthermore in this case study the influence and benefits of increasing the level of detail regarding the definition of quantity centres were examined.

The advantage of the MFCA, compared to conventional cost accounting, is the direct allocation of the costs also to the waste. An increasing level of detail in MFCA offers the opportunity to extract insights relevant to the respective target group. In this way, potential savings can be directly identified through a more detailed analysis and thus support process improvement. As a result, especially the economic effects of process changes can be demonstrated and the process understanding can be increased. This is necessary in order to be able to derive corresponding improvements from the MFCA, for example in terms of equipment. However, increasing the level of detail is associated with growing data requirements and thus higher temporal effort required for this method. This is difficult in early stages of process development, as some information is missing. Therefore, assumptions must be made, which in turn affect the validity of the data. In addition, the MFCA only indirectly takes the impact on the environment into account, so that a link with LCA is possible for a more detailed study of the ecological aspects. Overall, in the context of process development the advantages of increasing the level of detail in terms of the direct identification of perpetrators on the equipment level outweigh the expenses for the more extensive data collection.

Referenes

BMUB—Bundesministerium für Umwelt, Naturschutz, Bau und Reaktorsicherheit: Nachhaltige Entwicklung als Handlungsauftrag (2018). https://www.bmub.bund.de/themen/ nachhaltigkeit-internationales/nachhaltige-entwicklung/strategie-und-umsetzung/nachhaltigkeit-als-handlungsauftrag/

Hirzel, M., Gaida, I., Geiser, u.: Prozessmanagement in der Praxis. Springer Gabler Verlag, Wiesbaden (2013). ISBN 978-3-8349-4576-1

ISO 14051:2011–12: Environmental management—Material flow cost accounting—General framework

Klöpffer, W., Grahl, B.: Ökobilanz, 1st edn. WILEY-VCH Verlag GmbH & Co. KGaA, Weinheim (2009)

METI—Ministry of Economy, Trade and Industry: Material Flow Cost Accounting Case Samples, Tokyo (2010)

Schrack, D.: Nachhaltigkeitsorientierte Materialflusskostenrechnung. Springer Fachmedien, Wiesbaden (2016). ISBN 978-3-658-11301-8

Viere, T, Möller, A., Prox, M.: Materialflusskostenrechnung—ein Ansatz für die Identifizierung und Bewertung von Verbesserungen in der Ökobilanz. In: Feifel, S. et al. (ed.) Ökobilanzierung 2009—Ansätze und Weiterentwicklungen zur Operationalisierung von Nachhaltigkeit, conference transcript Ökobilanz-Werkstatt, Karlsruhe (2009)

von Hauff, M., Kleine, A.: Nachhaltige Entwicklung—Grundlagen und Umsetzung. Oldenbourg Verlag, München (2014). ISBN 987-3-486-85094-9

Wengerter, M., Scholl, S., Nieder, H.: Schneller, preiswerter, effizienter—kontinuierliche Bindemittelproduktion, Farbe und Lack, p. 10 (2017)

Wesche, M., Häberl, M., Kohnke, M., Scholl, S.: Ökologische Bewertung von Produktionsprozessen in Mehrproduktbatchanlagen, Chem. Ing. Tech. **87**(3) (2015). https://doi.org/10.1002/cite.201400086

Part IV
Case Studies

Development of a Functional Unit for a Product Service System: One Year of Varied Use of Clothing

Felix M. Piontek, Max Rehberger and Martin Müller

Abstract Product-service systems (PSS) are considered to provide environmental benefits compared to conventional consumption. The textile industry causes many negative environmental and social impacts. In our research project, we conduct a life-cycle assessment of a use-oriented PSS: The renting of casual clothing. In this article, we present the approach to develop our functional unit "One year of varied use of clothing" and that of the baseline-scenario.

Keywords Life-cycle assessment · Product-service systems · Clothing Functional unit

1 Introduction

The current state of the textile industry has major ecological, economic and social impacts. Especially the current "fast fashion"-trend causes increasing negative effects on the environment and the people employed in the textile industry who produce raw materials and clothes. The textile sector produces large quantities of solid waste and wastewater and consumes a lot of resources (Kozlowski et al. 2012). To produce raw materials, especially cotton fibers, genetically modified seeds (Kaur et al. 2013) and pesticides (EJF 2007) are widely used. Due to low prices and high competition, workers are exploited (Fletcher 2013).

Currently several ways to target these issues are suggested by different authors. Besides reduction of inputs and waste (Kozlowski et al. 2012) and sustainable design (e.g. Fletcher 2013) literature considers Product-service systems (PSS) as one way to reduce negative effects (Roy 2000; Baines et al. 2007). They are considered to be less

F. M. Piontek (✉) · M. Müller
Institute of Sustainable Corporate Management, Helmholtzstraße 18, 89081 Ulm, Germany
e-mail: felix.piontek@uni-ulm.de

M. Rehberger
Department of Business Chemistry at the Institute of Theoretical Chemistry, Helmholtzstraße 18, 89081 Ulm, Germany

© Springer Nature Switzerland AG 2019
L. Schebek et al. (eds.), *Progress in Life Cycle Assessment*, Sustainable
Production, Life Cycle Engineering and Management,
https://doi.org/10.1007/978-3-319-92237-9_11

harmful than conventional production and use patterns (Mont 2002), in case of the textile industry mass production of clothes and the current trend of "fast-fashion" (quick change of trends and offered styles to sell a lot of cheap and non-durable garments) (Fletcher 2013).

Product-Service Systems (PSS) are, according to the definition of Goedkoop et al., "a marketable set of products and services capable of jointly fulfilling a user's need" (Goedkoop et al. 1999). Tukker presented eight archetypical models of PSS in 2004 (Tukker 2004) and published a review on PSS eleven years later (Tukker 2015). He states that renting, a use-oriented PSS, can have significant benefits due to the more intensive use of goods. He also discusses limitations of PSS in a business-to-consumer context. The value of owning things instead of sharing them seems to be an important obstacle.

2 Business Model: Renting of Casual Wear

Based on interviews conducted with the founders of three start-ups in Germany, the business model shown in Fig. 1 is modeled (Apel 2016; Wilkening and Fendel 2016; Krichel 2016).

The arrow on the left and the grey "End-of-life"-process on the right represent the (conventional) value chain of the textile industry. End-of-life will be neglected in this work since it remains unclear whether the disposal after renting differs in any way from the disposal in conventional consumption. The white processes are the circular business model provided by the start-ups.

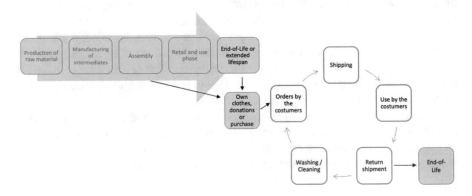

Fig. 1 Business model: Renting of casual wear

3 Proposed Method

Though many Life Cycle Assessment (LCA, ISO 2006) articles dealing with textiles exist, there are only a few which are dealing with PSS in the textile industry (Piontek and Müller 2018). While many articles analyzing the life cycle of t-shirts (Zhang et al. 2015), different fibres (Shen et al. 2010) or a new technology (Agnhage et al. 2017) use e.g. a number of garments or a certain amount of material as the functional unit, this approach does not seem feasible for a PSS providing the possibility for an alternative use pattern like renting. To model the divergent impacts of an alternative consumption pattern, a broader approach is needed. Castellani et al. conducted a study of avoided impacts of the activity of a second-hand shop which includes T-shirts and sweaters (Castellani et al. 2015).

Zamani et al. recently published a paper assessing the environmental impacts of a fashion library (Zamani et al. 2017). They focused on the prolonged service life of three types of garments in Sweden. We, in contrast, want to focus on the changed impacts of alternative clothing consumption of one consumer renting some of her clothes instead of buying them.

Therefore, we developed the functional unit "one year of varied use of clothing" by combining data from different sources. It represents the clothing consumption of one female consumer in Germany during one average year. We used data on purchase frequencies provided by a study by Spiegel Media (Spiegel QC 2015). Using data by Greenpeace (Greenpeace e.V. 2015) and of our research project (Geiger et al. 2017), we estimated the average period of use. Information on which types of clothes are rented have been provided by the start-ups (Apel 2017; Fendel 2017).

In Fig. 2 a schematic example of the functional unit of conventional consumption of clothes (purchase, use and disposal) is simulated. It shows a period of six seasons (one year n from spring to winter and two additional seasons (n − 1 and n + 1) before and after for reasons of illustration).

It is the schematic result of the combination of the previously mentioned data. We model an average year of consumption n of one person by combining new purchases, disposed clothes and the number of clothes owned.

Jacket 1 and *Jacket 2* are bought sometime before year n, are owned during year n and will be disposed sometime after year n. *Jacket 3* gets purchased in autumn of

	Conventional consumption					
	Winter_n-1	Spring_n	Summer_n	Autumn_n	Winter_n	Spring_n+1
Jacket 1	Owned					
Jacket 2	Owned					
Jacket 3				Purchased and owned		
Jacket 4	Owned and disposed					

Fig. 2 Schematic representation of the functional unit of conventional consumption

Renting of some clothes						
	Winter_n-1	Spring_n	Summer_n	Autumn_n	Winter_n	Spring_n+1
Jacket 1	Owned					
Jacket 2					Rented	
Jacket 3				Rented		
Jacket 4	Owned and disposed					

Fig. 3 Schematic representation of the functional unit of a renting-scenario

year n and will be disposed several years later. *Jacket 4* was bought several years before year n and gets disposed after summer of year n.

Figure 3 shows the same year n, but now some of the clothes are rented. *Jacket 1* remains owned as shown before, but now *Jacket 2* is rented during winter and *Jacket 3* during autumn of year n. *Jacket 4* remains owned by the user and disposed of after use.

The described procedure will allow us to compare the environmental impacts of conventional consumption and conventional consumption combined with renting of one person in one year. We are planning to allocate the environmental impact of production of each garment proportional to the seasons of usage or ownership during the same year. For our example shown in Figs. 2 and 3 we assume an average period of use of 4 years (16 seasons) for a jacket. Therefore, for conventional consumption we will take 12/64 of the total production impacts of the four jackets into account. 12/64 because of 4 seasons for *Jacket* 1, 4 seasons for *Jacket* 2, 2 seasons (autumn and winter) for *Jacket* 3 and 2 seasons (spring and summer) for *Jacket* 4. For our renting scenario, we only have to consider 8/64 of the production impacts following the same procedure. It must be considered, that renting can cause additional impacts for shipping and cleaning (Zamani et al. 2017).

4   Summary and Outlook

We presented a brief overview of the development of the functional unit "one year of varied use of clothing" which we will use to assess the changed environmental impacts of renting clothes instead of buying them. Assessing all the impacts for different kinds of clothes (including end-of-life) of both scenarios will allow us to make a statement on whether renting has benefits for the environmental impacts of one average person. It will be possible to use different baseline—as well as renting-scenarios to consider the divergent consumption patterns of different consumer groups. With this attributional method, we only assess the environmental impacts of one person, which is adequate, as the interviewed start-ups are currently niche companies and provide an offer for people interested in alternative consumption as well as people who want to try a new style modeled (Apel 2016; Wilkening and Fendel 2016; Krichel 2016). To understand the overall influence of renting, a different approach will be needed. We plan to simulate the business model shown in Fig. 1 for a fictional company providing renting of clothes on a bigger scale.

We also think that it will be possible to use this approach to conduct LCA-studies in different regions (using different input-data for the functional unit) or even apply it to other branches (e.g. car sharing or tool rental).

Acknowledgements We want to thank Samira Iran (TU Berlin), Markus Dimmer and Viktor R. Moritz Meissner (both Ulm University) for their input and support.

References

Agnhage, T., Perwuelz, A., Behary, N.: Towards sustainable *Rubia tinctorum L.* dyeing of woven fabric: how life cycle assessment can contribute. J. Clean. Prod. **141**, 1221–1230 (2017)

Apel, L. (Kleiderrebell): Personal communication. (2017, January 09)

Apel, L. (Kleiderrebell): Personal interviews together with Carolin Becker-Leifhold. (2016, June 13)

Baines, T.S., Lightfood, H.W., Evans, S., Neely, A., Greenough, R., Peppard, J., Roy, R., Shehab, E., Braganza, A., Tiwari, A., Alcok, J.R.: State-of-the-art in product-service systems. Proc. Inst. Mech. Eng. Part B J. Eng. Manuf. **221**(10), 1543–1552 (2007)

Castellani, et al.: Beyond the Throwaway Society: a life cycle-based assessment of the environmental benefit of reuse. Integr. Environ. Assess. Manage. **11**, 373–382 (2015)

EJF.: The deadly chemicals in cotton, environmental justice foundation in collaboration with pesticide action Network UK. London, UK (2007). ISBN No. 1-904523-10-2

Fendel, P. (Kleiderei): Personal communication. (2017, July 18)

Fletcher, K.: Sustainable fashion and textiles: design journeys. Routledge (2013)

Geiger, S.M., Iran, S., Müller, M.: Nachhaltiger Kleiderkonsum in Dietenheim. Ergebnisse einer repräsentativen Umfrage zum Kleiderkonsum in einer Kleinstadt im ländlichen Raum in Süddeutschland. Studienbericht. Universität Ulm (2017)

Goedkoop, M.J., Van Halen, C.J., Te Riele H.R., Rommens, P.J.: Product service systems, ecological and economic basics. Report for Dutch Ministries of environment (VROM) and economic affairs (EZ). **36**(1),1–122 (1999)

Greenpeace e.V.. Wegwerfware Kleidung. Repräsentative Greenpeace-Umfrage zu Kaufverhalten, Tragedauer und der Entsorgung von Mode. (2015)

Kaur, A., Kohli, R.K., Jaswal, P.S.: Genetically modified organisms: an Indian ethical dilemma. J. Agric. Environ. Ethics **26**(3), 621–628 (2013)

Kozlowski, A., Bardecki, M., Searcy, C.: Environmental impacts in the fashion industry: a life-cycle and stakeholder framework. J. Corp. Citizensh. **45** (2012)

Krichel, J. (Temporary Wardrobe): Personal interview together with Carolin Becker-Leifhold. (2016, May 25)

Mont, O.K.: Clarifying the concept of product–service system. J. Clean. Prod. **10**(3), 237–245 (2002)

Organización Internacional de Normalización: ISO 14044: Environmental management, life cycle assessment requirements and guidelines. ISO, Geneva, Switzerland (2006)

Piontek, F.M., Müller, M.: Literature reviews: life cycle assessment in the context of product-service systems and the textile industry, Procedia CRIP. **6**(1), 758–763 (2018)

Roy, R.: Sustainable product-service systems. Futures **32**(3), 289–299 (2000)

Shen, L., Worrell, E., Patel, M.K.: Environmental impact assessment of man-made cellulose fibres. Resour. Conserv. Recycl. **55**(2), 260–274 (2010)

Spiegel QC. Outfit 9.0 Zielgruppen—Marken—Medien. (2015)

Tukker, A.: Eight types of product–service system: eight ways to sustainability? Experiences from SusProNet. Bus. Strategy Environ. **13**(4), 246–260 (2004)

Tukker, A.: Product services for a resource-efficient and circular economy–a review. J. Clean. Prod. **97**, 76–91 (2015)

Wilkening, T., Fendel, P. (Kleiderei): Personal interview together with Carolin Becker-Leifhold. (2016, May 30)

Zamani, B., Sandin, G., Peters, G.M.: Life cycle assessment of clothing libraries: can collaborative consumption reduce the environmental impact of fast fashion? J. Clean. Prod. **162**, 1368–1375 (2017)

Zhang, Y., Liu, X., Xiao, R., Yuan, Z.: Life Cylce Assessment of cotton t-shirts in China. Int. J. Life Cycle Assess. **20**(7), 994–1004 (2015)

LCA in Process Development: Case Study of the OxFA-Process

Simon Rauch, Frank Piepenbreier, Dorothea Voss, Jakob Albert
and Martin Hartmann

Abstract In the context of environmental problems and resource depletion, the chemical industry is challenged with the substitution of fossil feedstock by renewable resources. To reduce the probability of unexpected environmental impacts by new processes using various sources of biomass, LCA methods can provide support for process development. The investigated OxFA-process, which converts biomass into formic acid, is a typical representative for these next generation processes. However, at an early development stage, data gaps have to be handled. In this study, scientific literature, process simulation and generic data have been used to identify the catalyst, compressed air and energy consumptions as those process components are the ones that significantly influence the environmental footprint of the whole process. Hereby starting points for an effective process optimization are shown and different scenarios for process implementation are compared. Furthermore, a first estimation of the competitiveness with the state-of-the-art process was attempted. Despite some data inaccuracies and required estimations, it is fair to assume that the OxFA-process is an ecological reasonable alternative.

Keywords Formic acid · OxFa process · Life cycle assessment · Biomass

1 Introduction

Limited supply of oil, natural gas and coal, as well as the environmental impact, caused by the combustion of fossil hydrocarbons, like global warming and the acidification of the oceans, calls for a substitution of fossil feedstocks. In the field of

S. Rauch · F. Piepenbreier · M. Hartmann (✉)
Erlangen Catalysis Resource Center, Friedrich-Alexander-Universität Erlangen-Nürnberg,
Egerlandstraße 3, 91058 Erlangen, Germany
e-mail: martin.hartmann@ecrc.uni-erlangen.de

D. Voss · J. Albert
Chair of Chemical Reaction Engineering, Friedrich-Alexander-Universität Erlangen-Nürnberg,
Egerlandstraße 3, 91058 Erlangen, Germany

© Springer Nature Switzerland AG 2019
L. Schebek et al. (eds.), *Progress in Life Cycle Assessment*, Sustainable
Production, Life Cycle Engineering and Management,
https://doi.org/10.1007/978-3-319-92237-9_12

green chemistry, biomass is seen as a promising renewable resource that can be converted into fuels or chemicals for further processing. However, chemical processes cannot simply be termed as "green" or "sustainable", because of the use of biomass (Gustafsson and Börjesson 2007). Environmental impacts induced by the consumption of (metal-based) catalysts, energy and other auxiliaries have to be assessed as well (Benavides et al. 2016). It is important to detect whether burdens of the process itself are reduced or simply shifted. Thus, for a holistic overview of environmental impacts of chemical processes, a Life Cycle Assessment (LCA) is recommended (Burgess and Brennan 2000; Tufvesson et al. 2013).

This study uses LCA methods and process simulation for a first estimation of environmental impacts of an innovative process on an early development stage based on current literature. This gives the chance to support further development, highlight chances to improve the environmental performance and evaluate different scenarios for the implementation of the process into industrial practice (Azapagic 1999). For a case study, the OxFA-process is applied as a typical representative for a process using biogenic feedstock, which aims at substituting a conventional process based on fossil resources.

Formic acid is used for animal feed preservation, textiles and leather processing, flue gas desulfurization and as intermediate in the pharmaceutical industry (2017; Reutemann and Kieczka 2005). For many applications, pure formic acid is not required and aqueous mixtures with a content of 60 wt% formic acid are used. The OxFA-process was developed to convert biomass into formic acid (FA), with carbon dioxide as sole side product. Equation 1 shows the simplified chemical reaction equation.

$$\text{Substrate} + O_2 \rightarrow HCOOH + CO_2 \qquad (1)$$

As substrate, a wide range of first, second and third generation biomasses can be used (Albert and Wasserscheid 2015). For this reason, Albert and Wasserscheid crafted a scenario for a decentralised chemical plant for the production of green fuel directly at a farm (Albert et al. 2016). The structure of the biomass employed affects the reaction rate, solubility, selectivity and conversion rate, which represent key process parameters that influence the environmental performance of the whole process. The reaction takes place at 90 °C in an aqueous solution. A higher temperature would increase the reaction rate, however, formic acid decomposes above 100 °C (Reutemann and Kieczka 2005). For the pulping of water-insoluble feedstock, *para*-toluenesulfonic acid (p-TSA) is recommended as an additive (Albert 2014). Figure 1 presents a simplified flowsheet of the OxFA-process.

A Keggin-type heteropoly acid with the molecular formula $H_8PV_5Mo_7O_{40}$ (HPA-5) currently shows the best catalytic performance in terms of activity and selectivity to formic acid (Albert 2014; Reichert and Albert 2017). The reaction proceeds in two steps: First, the homogeneous catalyst oxidizes the substrate. Thereby the catalyst is reduced. Afterwards, the catalyst has to be reoxidized using molecular oxygen at the lab-scale.

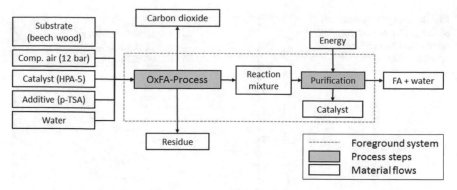

Fig. 1 Simplified flowsheet of the OxFA-Process

To avoid a reaction rate reduction, the reoxidation of the catalyst has to be faster than the oxidation of the biomass. The former is influenced by the partial pressure of oxygen in the reactor, while the latter is governed by the chemical structure of the substrate. In the literature an oxygen pressure of 20 bar is suggested (Albert and Wasserscheid 2015). The use of complex biomass results in a slower reaction rate, thus a system pressure of 12 bar is assumed to be sufficient. On industrial scale, compressed air is used instead of pure oxygen, because of economic and ecological reason.

After the reaction, the reaction mixture is purified to a saleable product via distillation. The formation of a high-boiling azeotrope, close-boiling components and the thermal sensitivity of formic acid (FA) complicate the separation via rectification. First, a water rich phase is removed as low boiler overhead until the desired concentration of FA is attained. In a second step, the product is separated from the residue and the catalyst, which remain in the bottom and will be recycled subsequently.

2 Materials and Methods

Goal and Scope: The goal of this study is providing support for development of the OxFA-Process by identifying significant parameters that influence the environmental performance. Therefore the target audience are scientists improving this process. An attributional LCI modelling framework is used.

The functional unit is set to 1 kg of pure formic acid (FA). The formic acid leaves the OxFA-process as aqueous solution with a concentration of 60 wt% FA, because this is a good trade-off between product purity and energy consumption (Fig. 2). In Fig. 1 the foreground system of this study is shown. For inputs from the background system, generic data is taken from the ecoinvent database v 3.3 using the cut-off approach. The residue is seen as biowaste and is treaded in a municipal incineration

Fig. 2 Energy consumption of the purification depending on the feed and product concentration

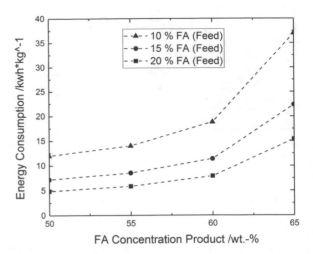

plant. Transport of the feedstock is included, if it is necessary for the particular scenario.

Life Cycle Inventory: Relevant data for the LCI are the amount of consumed oxygen and substrate, the amount of residue in the case of incomplete conversion, as well as the loss of additive and catalyst. Furthermore, the energy consumption of the purification step is taken into account. The required equipment and emissions of the OxFA-process itself into air, water and ground are not included in this inventory, because at the current development stage their uncertainty is too high.

Biomass and oxygen consumption are calculated according to the conversion and structural composition of the substrate biomass. Therefore it is assumed that the residue after the reaction has the same composition as the substrate and that oxygen is consumed in a stoichiometric ratio. Accordingly, a high specific yield of formic acid results in lower substrate demand per formed kg formic acid and consequently less residue. The selectivity of the formic acid formation influences the oxygen consumption, because the less carbon dioxide is formed the less oxygen is stoichiometrically consumed with regard to the functional unit of 1 kg FA. From the wide range of possible biomass feedstocks, beech wood from sustainable forestry was chosen as substrate in this study. With respect to the carbon atoms, the yield of formic acid is 29% (Albert 2014). In the literature, no details regarding the loss of catalyst and additive are provided. Therefore, the assumption is made that 10 g of additive and 1 g catalyst are consumed per kilogram of formic acid. This assumption has to be checked in a long-time study.

The catalyst is synthesized from phosphoric acid, hydrogen peroxide, molybdenum(VI)-oxide and vanadium(V)-oxide (Odyakov and Zhizhina 2008). Vanadium(V)-oxide is not included in the ecoinvent-database. For this reason a dataset for the production of vanadium(V)-oxide from vanadium-slag by the roast-leach-process was created. Vanadium-slag, a by-product of iron mining, is converted to vanadium(V)-oxide with sodium carbonate and ammonium sulfate (Goso et al.

Component	Amount/kg FA
Substrate (beech wood)	2.03 kg
Residue	0.96 kg
Comp. air (12 bar)	3.87 kg
Catalyst (HPA-5)	0.001 g
Additive (p-TSA)	0.01 kg
Energy	11.4 kWh

Table 1 Inventory data of the OxFA-process related to1 kg FA

2016; Nkosi 2017). The allocation of the environmental impacts of iron mining is done by mass, because the vanadium price is subject to strong fluctuations (Mokalyk and Alfantazi 2013).

3 Results and Discussion

Formic acid is separated from excess water, residue and catalyst via batch-distillation. Since the formic acid concentration neither in the reaction mixture nor in the desired product is known, the energy consumption of the distillation of 1 kg FA in aqueous solution is calculated for different formic acid concentrations with Aspen Batch Modeler V9 (Fig. 2). Energy consumption is rising with decreasing feed (reaction mixture) concentration and increasing product concentration of the distillation. Close to the azeotrope, the energy consumption is increasing almost exponentially. For our study, a formic acid concentration of 15 wt% in the reaction mixture and 60 wt% in the product are assumed, which results in an energy consumption of 11.4 kWh/kg FA. Inventory data are collected in Table 1.

Concerning energy supply, three different scenarios are evaluated. In the scenario "Market", electricity is received from the German power market, while the scenario "PV" makes use of electricity from a photovoltaic system.

In addition, the utilization of thermal energy from a cogeneration unit is investigated in the scenario "Heat". Furthermore a transport distance of 50 km for the feedstock is taken into account. The idea of the latter scenario is the centralized processing at an industrial area with existing heat network, while the first two scenarios represent the idea of a decentralized biomass processing at a farm (Albert et al. 2016).

Life Cycle Impact Assessment: With the inventory data described above, a first impact assessment was made. As impact assessment method ReCiPe 2008 is used with the hierarchical value choice (Goedkoop et al. 2008). The results for the five midpoint indicators Global Warming Potential (GWP), Ozone Depletion Potential (ODP), Metal Depletion Potential (MDP), Human Toxicity Potential (HTP) and Freshwater Eutrophication Potential (FEP) are shown in Fig. 3 for the three different

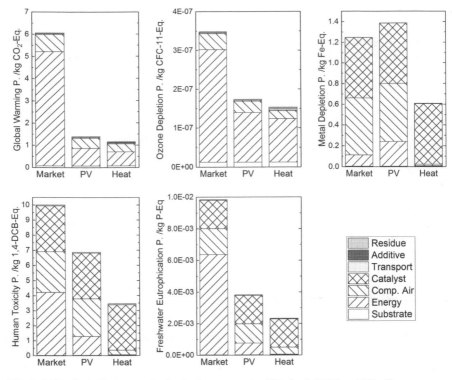

Fig. 3 Midpoint indicator results for the three scenarios "Market", "PV" and "Heat"

energy scenarios. Carbon dioxide emissions from biogenic origin are considered as neutral with respect to GWP.

Catalyst synthesis and consumption, compressed air production and energy consumption for the purification step are the key contributors to the environmental footprint according to this evaluation. Therefore, process improvements concerning these elements will have the largest benefits. Additive and substrate consumption as well as the treatment of residue are of little relevance. The same results for biomass transport in the scenario "Heat". However, changing the energy source can significantly reduce the environmental impact (Fig. 3).

Electricity and heat consumption for the purification step and compressed air production predominantly contribute to GWP. Emissions from lignite- and coal-fired power stations are main contributors to the GDP of the scenario "Market". A similar result is found for the Ozone Depletion Potential, which is by far a factor of two higher for the "Market" scenario compared to "PV" and "Heat".

Compressed air production strongly contributes to the Metal Depletion Potential in the decentralized scenarios "Market" and "PV", but not in the scenarios "Heat". This contribution is caused by the wear of the compressor. In the centralized scenario "Heat" a compressor with a higher capacity is used, which reduces the wear in relation

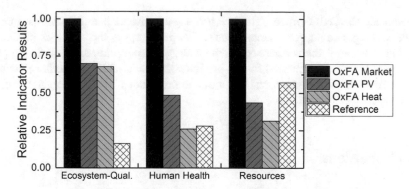

Fig. 4 Comparison of the three OxFA-scenarios and the methyl formate route at the endpoint level

to the amount of compressed air (Steiner and Frischknecht 2007). Molybdenum and Vanadium, which are used for the synthesis of the catalyst also contribute to the MDP. In a similar way, they affect Human Toxicity Potential. In particular mining operations contribute to negative effects on human health and freshwater eutrophication.

In the next step, the potential environmental impacts of the three scenarios of the OxFA-process are compared with the conventional process for the production of formic acid via the methyl formate route. The feedstock for this process is carbon monoxide and water (Reutemann and Kieczka 2005). Because carbon monoxide is mainly produced from coal or gas, this process depends on fossil resources. Inventory data is taken from Sutter (2007). Not included in the evaluation of the reference process is the compression of the gases to 45 bar and the consumption of the typically employed catalyst (sodium methoxide). Both factors which show a significant influence on the environmental balance of the OxFA-process. However, a formic acid concentration of up to 98 wt% is reached.

In Fig. 4, the results for the three endpoint indicators human health, ecosystem-quality and resources for the three OxFA-scenarios and the reference process are compared. Endpoint indicators are used, because all environmental impacts should be considered, when both techniques are compared.

Regarding ecosystem-quality, the impacts of the different OxFA-process scenarios are significantly higher compared to those of the reference process. The ecosystem-quality is mainly affected by provision of the substrate (beech wood), which requires a large area for cultivation. Therefore, using waste biomass for the OxFA-process, would reduce the environmental impact significantly. However, waste biomass is only available in a limited amount and increased competition for its use also has to be considered.

Impacts on human health of the scenario "Heat" are comparable with those of the reference process, while the impact on human health of the other scenarios is somewhat higher. Thus, it has to be concluded that the competitiveness of the OxFA-process depends on the energy source chosen. The same applies for resource depletion. Using heat or solar power reduces resource consumption. Under these circumstances, the OxFA-process shows a better performance than the state-of-the-art process. Suggested improvements of the OxFA-process, viz. the in situ extraction of FA with 1-hexanol can further reduce the energy consumption and enable the production of highly concentrated formic acid (Reichert et al. 2015). Another potential improvement is the heterogenization of the homogeneous HPA-5 catalyst, which is currently under investigation.

4 Conclusions

In this study, an evaluation of the environmental footprint of a chemical process at an early stage of development was attempted. Therefore, assumptions had to be made regarding the consumption of auxiliaries like additive and catalyst. The amount of oxygen used had to be calculated based on the assumed stoichiometry laws and, thus, describes the technical optimum, while emissions of the OxFA-process itself were not taken into consideration. Nevertheless, a first estimation with respect to the key contributions to the environmental footprint was made. In our case, the catalyst preparation, energy consumption and compressed air production are the most significant contributors. Improvements which focus on these aspects seem to be the most promising approaches for further development. Finally, it can be assumed that the OxFA-process can compete with the state-of-the-art technology, when the energy supply is based on renewable resources such as photovoltaics or heat cogeneration.

References

Albert, J.: Chemische Wertschöpfung aus Biomasse mittels selektiver katalytischer Oxidation zu Ameisensäure (FA)—der Erlanger OxFA-Prozess, Dissertation (2014)

Albert, J., Wasserscheid, P.: Expanding the scope of biogenic substrates for the selective production of formic acid fromwater-insoluble and wet waste biomass. Green Chem. **17**, 5164–5171 (2015)

Albert, J., Jess, A., Kern, C., Pöhlmann, F., Glowienka, K., Wasserscheid, P.: formic acid-based fischer–tropsch synthesis for green fuel production from wet waste biomass and renewable excess energy. ACS Sustain. Chem. Eng. **4**, 5078–5086 (2016)

Azapagic, A.: Life Cycle assessment and its applications to process selection, design and optimisation. Chem. Eng. J. **73**, 1–21 (1999)

Benavides, P.T., Cronauer, D.C., Adom, F., Wang, Z., Dunn, J.B.: The influence of catalysts on biofuel life cycle analysis (LCA). Sustain. Mater. Technol. **11**, 53–59 (2016)

Burgess, A.A., Brennan, D.J.: Application of life cycle assessment to chemical processes. Chem. Eng. Sci. **56**, 2589–2604 (2000)

http://www.intermediates.basf.com/chemicals/formic-acid/index Accessed: 23 Nov 2017

Goedkoop, M., Heijungs, R., Huijbregts, M., Schryver, A., Struijs, J., van Zelm, R.: ReCiPe 2008—A life cycle impact assessment method which comprises harmonised category indicators at the midpoint and the endpoint level (2013)

Goso, X.C., Lagendijk, H., Erwee M., Khosa, G.: Indicative vanadium deportment in the processing of Titaniferous Magnetite by the Roast-Leach and Electric Furnace Smelting Processes, Hydrometallurgy Conference (2016)

Gustafsson, L.M., Börjesson, P.: Life cycle assessment in green chemistry. A comparison of various industrial wood surface coatings. Int. J. LCA **12**, 151–159 (2007)

Mokalyk, R.R., Alfantazi, A.M.: Processing of vanadium: a review. Miner. Eng. **16**, 793–805 (2013)

Nkosi, S., Dire, P., Nyambeni, N., Goso, X.C.: A comparative study of vanadium recovery from titaniferous magnetite using salt, sulphate, and soda ash roast-leach processes. In: 3rd Young Professionals Conference (2017)

Odyakov, V.F., Zhizhina, E.G.: A novel method of the synthesis of molybdovanadophosphoric heteropoly acid solutions. React. Kinet. Catal. Lett. **95**, 21–28 (2008)

Reichert, J., Albert, J.: Detailed kinetic investigations on the selective oxidation of biomass to formic acid (OxFA Process) using model substrates and real biomass. ACS Sustain. Chem. Eng. **5**, 7383–7392 (2017)

Reichert, J., Brunner, B., Jess, A., Wasserscheid, P., Albert, J.: Biomass oxidation to formic acid in aqueous media using polyoxometalate catalyst—boosting FA selectivity by in-situ extraction. Energy Environ. Sci. **8**, 2985–2990 (2015)

Reutemann, W., Kieczka, H.: Formic Acid, in Ullmann's Encyclopedia of Industrial Chemistry (2005)

Steiner, R., Frischknecht, R.: Metals processing and compressed air supply, ecoinvent report No. 23. Swiss Centre for Life Cycle Inventories, Dübendorf (2007)

Sutter, J.: Life cycle inventories of petrochemical solvents, ecoinvent report No. 22. Swiss Centre for Life Cycle Inventories, Dübendorf (2007)

Tufvesson, L.M., Tufvesson, P., Woodley, J.M., Börjesson, P.: Life cycle assessment in green chemistry: overview of key parameters and methodological concerns. Int. J. LCA **18**, 431–444 (2013)

Using Energy System Modelling Results for Assessing the Emission Effect of Vehicle-to-Grid for Peak Shaving

Anika Regett, Constanze Kranner, Sebastian Fischhaber and Felix Böing

Abstract An understanding of the surrounding energy system is necessary when using the system expansion approach for an emission assessment of sharing and reuse concepts for electric vehicle batteries. In this context it is shown that existing energy system modelling results can be used for assessing the marginal effect of a load change on CO_2 emissions for electricity using either the electricity mix or the marginal power plant method. From the company's perspective a load management with vehicle-to-grid for peak shaving leads to a decrease of electricity costs due to a reduced demand charge, but percentage changes in direct CO_2 emissions for electricity demand are small. The impact of vehicle-to-grid for peak shaving to reduce greenhouse gas emissions associated with the traction batteries' production strongly depends on the reference technology and is significant in case a diesel generator is substituted.

Keywords Energy system modelling · Electric vehicles · Charging management Vehicle-to-Grid · Emission assessment · Circular economy

1 Introduction

While reducing the demand for fossil fuels the energy transition requires new technologies, e.g. wind power plants and battery storage systems, many of which come with an increasing demand for critical materials such as cobalt and rare earth metals (Blagoeva et al. 2016). The criticality of a material can be caused by supply and/or environmental risks associated with its provision (Regett and Fischhaber 2017). Therefore in order to counteract the occurrence of new resource risks, measures are needed to reduce resource criticality of key technologies for future energy supply. The circular economy (CE) is often proposed as a means to reduce the

A. Regett (✉) · C. Kranner · S. Fischhaber · F. Böing
Forschungsstelle für Energiewirtschaft (FfE) e.V, Am Blütenanger 71, 80995 Munich, Germany
e-mail: aregett@ffe.de

© Springer Nature Switzerland AG 2019
L. Schebek et al. (eds.), *Progress in Life Cycle Assessment*, Sustainable Production, Life Cycle Engineering and Management,
https://doi.org/10.1007/978-3-319-92237-9_13

environmental impact of raw materials and to create new opportunities for value creation (European Commission 2017). Thus, there is a need to assess the potential of innovative CE approaches such as sharing and reuse to reduce the material-related environmental and supply risks of energy technologies.

In this context, the following main methodological challenges of assessing sharing and reuse concepts of electric vehicle (EV) batteries have been identified based on a literature review covering Life Cycle Assessment (LCA) studies of vehicle-to-grid (V2G) applications (e.g. Hoehne and Chester 2016; Yang et al. 2016) and stationary second-life applications (e.g. Kim et al. 2015; Richa et al. 2015) respectively:

- The battery lifetime and the provided storage function are influenced by the battery aging process, which is dependent on various factors such as the load profile and the battery's state-of-charge.
- Due to a longer lifespan in a second-life-application, the future development of waste management processes needs to be considered.
- The emissions need to be allocated to different battery functions, for example a mobility service (in pkm) and an energy system service (in kW or kWh).
- When choosing the system expansion approach for dealing with the allocation problem, an understanding of the energy system is needed in order to quantify the emissions of the substituted technologies.

Addressing the latter challenge, this paper aims to demonstrate how energy system modelling results can be used to assess the effects of V2G for a peak shaving application on electricity-related CO_2 emissions.

2 Methods

As an exemplary sharing concept, a bidirectional charging management for industrial peak shaving in an office building is analysed. In this case the power and storage capacity of the EV batteries is shared between the car owners and the company.

The analysed load profile originally stems from a real office building of a software company with around 100 employees, which is complemented by the load for directly charging the EV fleet. The load profile is characterized by a maximum annual peak load of 141.3 kW. The EV charging process is responsible for the load peaks in the morning. The load peaks in summer are induced by air conditioning, for which the load shifting potential is limited.

Therefore in order to decrease the peak load and thereby reduce the demand charge for electricity, a load management strategy with V2G is examined. In the V2G scenario, the EV fleet consists of 3 vehicles being available between 7 am and 6 pm. Each EV battery system is characterised by a 20 kWh storage capacity, 22 kW charging and discharging power, as well as a system efficiency of 85%. Under these circumstances, the maximum peak load can be reduced to 80.4 kW. This is done by shifting the charging process and by deliberately charging and discharging the EV batteries, whilst ensuring that the state-of-charge at the end of the day is equal to the case without peak shaving. Next to the comparison with the unsmoothed load profile, the V2G scenario is further compared to the use of a diesel generator, which generates electricity every time the load boundary of 80.4 kW is exceeded. It is assumed that the diesel generator with an efficiency of 34% is purchased specifically for the peak load management application.

In order to assess the emission effect of V2G for industrial peak shaving, the changes in direct energy-related CO_2 emissions for electricity generation, due to the change in load profile, are quantified. For the comparison with the diesel scenario the emission changes due to the change in load profile and the saved CO_2 emissions from diesel combustion are accounted for. Thus, in both cases only the effects on electricity-related CO_2 emissions in the operation phase are considered.

The CO_2 emissions associated with the increase or decrease of electricity generation can be quantified using either the electricity mix method or the marginal power plant method as described in (Conrad et al. 2017). For the mix method, the CO_2 emission factor of electricity is calculated by weighting the specific emissions of each power plant with its share in total electricity generation every hour. For the marginal method, the hourly emission factor of electricity equals the specific emission factor of the marginal power plant. The marginal power plant is the last one to be dispatched every hour, and therefore increases or decreases electricity generation in case of a marginal change of load.

An emission tool has been developed which uses energy system modelling results, load profiles from a peak shaving model, as well as fuel-related emission factors as inputs, so as to calculate the hourly emission factor of electricity with both the mix and marginal methods (see Fig. 1).

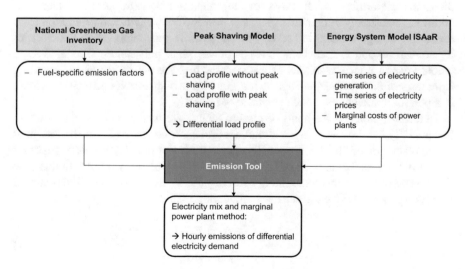

Fig. 1 Structure of the tool for calculating electricity-related emissions

The fuel-specific emission factors are taken from the national greenhouse gas inventory (Umweltbundesamt 2016) and entail only direct CO_2 emissions from fossil fuel combustion. This means for renewable energy the emission factor amounts to zero.

The load profiles needed for the assessment of the V2G scenario are calculated using a simple peak shaving model. Based on the company's load profile, including an uncontrolled charging process of the selected EV fleet, a smoothed load profile due to load shifting and V2G is calculated. For this purpose the EV fleet is considered as a storage system. The model derives the charging and discharging power of the storage system in order to keep the total load below a preset limit, while considering the battery parameters described above. The load limit and the number of EVs are chosen in a way that an almost optimal configuration for peak shaving of the analysed load profile can be achieved. For further calculations, a differential load profile is derived by subtracting the load profile for the scenarios "no peak shaving" and "diesel generator" from the modelled load profile of the V2G scenario.

Energy system modelling results, namely the time series of electricity generation and prices, as well as the marginal costs of power plants, are needed for calculating the emission factors for electricity in the emission tool. This input data is derived from simulation results for a scenario of the German energy system in 2030 which has been developed in the project "Merit Order Netz-Ausbau (MONA) 2030" (Regett et al. 2017) and is characterised by a share of 61% renewable energy. The simulation takes place in the energy system model ISAaR "Integrated Simulation Model for Planning the Operation and Expansion of Power Plants with Regionalisation", which is described in more detail in Pellinger et al. (2016) and Böing et al. (2017). One of the models' main features is the capability to perform an electricity market simulation for Europe. While European neighbours are modeled in a simplified way, the German-Austrian market zone is accurately represented. The power plants' dispatch is determined by the costs for fuel, operation and emission certificates. In addition, heat supply by combined heat and power (CHP) plants and generation from renewable energy systems are taken into account. In this composition the electricity generation and fuel consumption per power plant unit can be simulated for a whole year in hourly resolution. Strictly speaking, the electricity prices resulting from the simulation constitute hourly marginal costs of electricity generation, which in a perfect market correspond to electricity prices.

In the emission tool the hourly emission factors are then calculated according to Formula (1) for the electricity mix and according to Formula (2) for the marginal power plant method. The last term in the numerator in Formula (1) is included in order to allocate the total emissions of CHP plants to their electricity output. In this case the method of the International Energy Agency (IEA), as described in (International Energy Agency 2005), is chosen as an allocation method.

$$emf_{el,mix}(h) = \frac{\sum_{pp}\left(emf_f \times F_{pp}(h) \times \frac{E_{el,pp}(h)}{E_{out,pp}(h)}\right)}{\sum_{pp}\left(E_{el,pp}(h)\right)} \tag{1}$$

$emf_{el, mix}$ specific emission factor of electricity mix in kg CO_2/kWh
emf_f specific emission factor of fuel in kg CO_2/kWh
F_{pp} fuel consumption per power plant (pp) in kWh
$E_{el, pp}$ electrical generation per power plant (pp) in kWh
$E_{out, pp}$ total electrical and thermal generation per power plant (pp) in kWh.

As the last power plant sets the price in the day-ahead electricity market, the time series of electricity prices is matched with the marginal costs of each power plant in order to determine the marginal power plant for Formula (2). Should there be no exact match, the power plant with the next larger marginal costs is chosen. Because of the additional thermal output, CHP power plants are not defined as marginal.

$$emf_{el,m}(h) = emf_f \times \frac{F_{mpp}(h)}{E_{el,mpp}(h)} \tag{2}$$

$emf_{el, m}$ specific emission factor of marginal electricity demand in kg CO_2/kWh
emf_f specific emission factor of fuel in kg CO_2/kWh
F_{mpp} fuel consumption of marginal power plant (mpp) in kWh
$E_{el, mpp}$ electrical generation of marginal power plant (mpp) in kWh.

Finally, the calculated time series of emission factors is multiplied with the differential load profile so as to determine the emissions of the change in electricity demand due to V2G for peak shaving. This is done both for the case without peak shaving as well as for the peak shaving scenario with a diesel generator. Additionally, for the diesel generator scenario the emissions from diesel combustion are subtracted.

3 Results and Discussion

When comparing V2G for peak shaving to the load profile without peak shaving in 2030 a slight annual reduction of CO_2 emissions for electricity of 38 kg/a for the mix method and a slight increase of 33 kg/a for the marginal method is observed. When attributing the emission savings of the V2G sharing concept to the traction battery the emissions for battery production can be reduced by around 6% for the mix method. This holds true for greenhouse gas emissions for battery production of around 100 kg CO-eq. per kWh battery capacity (Regett 2018) and a life time of electric vehicles of 10 years (Thielmann et al. 2017).

The comparison of V2G for peak shaving with the diesel generator scenario shows that the emission effect is strongly dependent on the substituted technology. The large emission savings of 2118 kg/a for the mix and 1355 kg/a for the marginal method are mainly due to the saved emissions from diesel combustion for smoothening the morning peaks induced by the EV fleet. If V2G substitutes a stationary battery with the same parameters as the EV fleet, the emission savings would amount to 326 kg/a

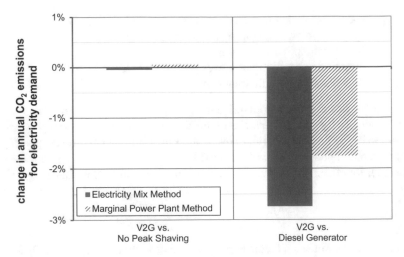

Fig. 2 Change in annual CO_2 emissions for electricity demand due to V2G for peak shaving compared to no peak shaving (left) and peak shaving by a diesel generator (right)

for the marginal and 161 kg/a for the mix method. These savings can be attributed to the additional losses for the stationary battery when shaving the load peaks from the EV charging process.

If an EV charging management is already implemented in the reference case so as prevent additional load peaks, the emissions savings would decrease. In case not only the operation phase, but also the prevented production of the substituted technology is accounted for, the emission savings would increase.

As can be seen from Fig. 2 the impact of peak load management on the total annual emissions for electricity demand of the analysed company are in both reference cases and methods negligible, with values lying in the range of model and data uncertainties.

Considering that in case of a peak load management application the amounts of electricity shaved or shifted are low (around 2.5% of the total annual electricity demand), a larger change in annual emissions can be expected for storage applications characterised by a greater number of charging and discharging cycles such as primary control reserve.

Furthermore, Fig. 2 shows that the comparison of V2G with the diesel scenario leads to a larger change in annual emissions than the comparison with the unsmoothed load profile. This can be explained by a larger difference in emission factors of electricity from the diesel generator (782 g/kWh) and electricity from the grid, compared to the difference in hourly emission factors of electricity from the grid due to load shifting.

For a better understanding of the difference between the marginal and the mix methods, the differential load profiles, as well as the specific emission factors of electricity for both methods, are shown in Fig. 3. The depicted time period is an exemplary day close to the average day of the year.

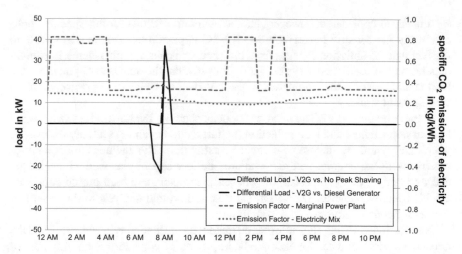

Fig. 3 Differential load (left axis) and emission factor of electricity for the marginal and the mix method (right axis) for an exemplary day

In general, the hourly emission factor of the electricity mix is, in about 85% of the year, lower than the hourly emission factor of the marginal power plant which is mostly gas- or coal-fired. However, the differences between the two methods vary over the day, meaning the impact of the chosen method depends on the analysed load profile.

Looking at the differential load profile for the V2G scenario compared to the unsmoothed load profile, it can be seen that a large share of the charging process is moved from 7 am to 8 am. As the emission factors for these two hours are very similar for both the mix and the marginal methods, neither a large change in annual emissions nor a large difference between the two methods can be observed. In case of a load shift the hourly emission factors at times of negative and positive differential load determine whether the emissions are equalled out. For the mix method the change in annual emissions would, for example, increase if the load is shifted between day and night as the differences in hourly emission factors are larger.

The differential load profile for V2G compared to the diesel scenario shows an increased electricity demand from the grid at 8 am due to load shifting, which is not required in the diesel scenario because the peak load is provided by the generator. For the diesel scenario the difference between the two methods is larger because the emissions for electricity from the diesel generator stay the same, while in the respective hours the reduced emissions from the grid are larger for the marginal than for the mix method.

Overall the results indicate that from the company's perspective the percentage changes in CO_2 emissions for electricity demand due to load management with V2G for peak shaving are low. In contrast, for an annual demand charge of 80 €/kW the peak load application leads to significant cost savings of around 5000 € per year.

It needs to considered, however, that the quantative results for this specific application are not directly transferable to all V2G applications. This is attributable to the fact that the emission changes from bidirectional charging management strongly depend on the used energy system modelling results, the analysed application, the load characteristics, the parameters of the EV fleet as well as the precise charging management strategy.

In the investigated V2G application for peak shaving, the sensitivities with regard to emission savings and load reduction are large and depend essentially on factors, which are specific to the use case under investigation. This includes the number of electric vehicles, attendance time, available battery capacity as well as level, time and duration of the load peaks. The chosen parameters and load profile constitute a V2G-friendly scenario. Thus, in a real case a thorough analysis of the described factors is needed before implementation.

Also the load shifting algorithm has an impact on the load shaving potential and the emission savings and can for example be improved by considering monthly instead of yearly peak loads. But considering that in the worst case the load reduction effect can amount to zero in case of a mismatch between load peaks and EV presence the sensitivity of the described use case-specific factors are more substantial.

4 Conclusions

Even though the exact impact of bidirectional charging management applications on CO_2 emission for electricity is determined by various factors, it can be noted that the change in emissions depends on the additional energy demand due to battery losses and the temporal shift of the load leading to different emission factors for electricity demand.

While peak load management leads to a decrease of electricity costs due to a reduced demand charge, in the analysed scenario a small change in CO_2 emissions was observed when compared to the total emissions for the company's electricity demand. This can be attributed to the small amounts of electricity affected by peak load management compared to the total electricity demand and the small difference between the hourly emission factors due to load shifting.

When attributing the emission savings from the V2G sharing application to the production of traction batteries it can be seen that the savings strongly depend on the substituted technology. While the impact is small for the comparison with the unsmoothed load profile, the emissions savings due to V2G can even exceed the emissions for battery production in case a diesel generator is substituted.

It is demonstrated that existing energy system modelling results can be used for assessing the environmental effects of a change of load using either the electricity mix or the marginal power plant method. With the marginal power plant method an approach is presented which considers system effects, but doesn't necessitate additional simulation runs. While this method is valid for assessing a marginal change, for the assessment of a large scale change in electricity demand a new simulation run of the energy system is necessary. These methodological aspects are further investigated in the research project "Dynamis" dealing with an assessment of cross-sector CO_2 abatement measures (Regett et al. 2017).

Ackknowledgements This research was supported by the "Stiftung Energieforschung Baden-Württemberg" and the "Hans und Klementia Langmatzstiftung".

References

Blagoeva, D.T., et al.: Assessment of potential bottlenecks along the materials supply chain for the future deployment of low-carbon energy and transport technologies in the EU—wind power, photovoltaic and electric vehicles technologies, time frame: 2015–2030. European Union, Luxembourg (2016)

Böing, F., Bruckmeier, A., Murmann, A., Pellinger, C., Kern, T.: Relieving the German Transmission Grid with Regulated Wind Power Development. In: 15th IAEE European Conference, Vienna, IAEE (2017)

Conrad, J., et al.: Evolution und Vergleich der CO_2-Bewertungsmethoden von Wärmepumpen. In: 3, Dialogplattform Power to Heat. Berlin, Energietechnische Gesellschaft ETG, VDE (2017)

European Commission: Report from the Commission on the implementation of the Circular Economy Action Plan. European Commission, Brussels (2017)

Hoehne, C.G., Chester, M.V.: Optimizing plug-in electric vehicle and vehicle-to-grid charge scheduling to minimize carbon emissions. Energy **115**, 646–657 (2016)

International Energy Agency: Energy Statistics Manual. International Energy Agency, Paris (2005)

Kim, D., et al.: Quantifizierung des Umweltnutzens von gebrauchten Batterien aus Elektrofahrzeugen als gebäudeintegrierte 2nd-Life-Stromspeichersysteme. Bauphysik **37**(4), 213–222 (2015)

Pellinger, C., Schmid, T., et al.: Merit Order der Energiespeicherung im Jahr 2030—Hauptbericht. Forschungsstelle für Energiewirtschaft e.V, Munich (2016)

Regett, A.: Ressourcensicht auf die Energiezukunft—2. Zwischenbericht. München: Forschungsstelle für Energiewirtschaft e.V. (2018)

Regett, A., Fischhaber, S.: Reduction of Critical Resource Consumption through Second Life Applications of Lithium Ion Traction Batteries. In: 10, Internationale Energiewirtschaftstagung (IEWT). Vienna, TU Wien (2017)

Regett, A., Zeiselmair, A., et al.: Merit Order Netz-Ausbau 2030—Teilbericht 1: Szenario-Analyse. Forschungsstelle für Energiewirtschaft e.V, Munich (2017a)

Regett, A., Conrad, J., Fattler, S.: Das Verbundprojekt Dynamis—Dynamische und intersektorale Maßnahmenbewertung zur kosteneffizienten Dekarbonisierung des Energiesystems. BWK **69**(1/2), 58 (2017b)

Richa, K., et al.: Environmental trade-offs across cascading lihtium-ion battery life cycles. Int. J. Life Cycle Assess. **22**(1), 66–81 (2015)

Thielmann, A., et al.: Energiespeicher-Roadmap (Update 2017)—Hochenergie-Batterien 2030 + und Perspektiven zukünftiger Batterietechnologien. Fraunhofer-Institut für System- und Innovationsforschung ISI, Karlsruhe (2017)

Umweltbundesamt.: Berichterstattung unter der Klimarahmenkonvention der Vereinten Nationen und dem Kyoto-Protokoll 2016—Nationaler Inventarbericht zum Deutschen Treibhausgasinventar 1990–2014. Dessau-Roßlau, Umweltbundesamt (2016)

Yang, Z., et al.: Vehicle to Grid regulation services of electric delivery trucks: Economic and environmental benefit analysis. Appl. Energy **170**, 161–175 (2016)

Assessment of Social Impacts Along the Value Chain of Automation Technology Components Using the LCWE Method

Friederike Schlegl, Mercedes Barkmeyer, Alexander Kaluza, Eva Knüpffer and Stefan Albrecht

Abstract *Purpose* During the last years, the sustainability of products has gained importance. The established life cycle assessment (LCA) methodology focuses on the evaluation of environmental impacts of products. Academic research extends this approach to social impacts towards establishing a social life cycle assessment (S-LCA). The life cycle working environment (LCWE) method offers one option to include social impacts in LCA. This paper aims at creating a LCWE procedure for company purposes. Automation technology components serve as a case study. *Methods* LCWE helps to assess the social impact on humans along a products life cycle based on statistical data and the possibility to integrate primary data. The methods focus is on working conditions in upstream and manufacturing activities. LCWE is based on the energy and material flows used in LCA study's. The results of the method are social profiles of single processes or products. The allocation to social profiles is performed through shares of value-added costs within the life cycle processes of a product. *Results and discussion* First, a procedure that enables the integration of the LCWE method into companies is presented. Within the procedure, it is possible to integrate corporate data as well as companies aims. The results of

F. Schlegl (✉)
Department of Life Cycle Engineering (GaBi), Institute for Acoustics and Building Physics (IABP), University of Stuttgart, Wankelstraße 5, 70563 Stuttgart, Germany
e-mail: friederike.schlegl@iabp.uni-stuttgart.de

M. Barkmeyer
Festo AG & Co. KG, Ruiter Straße 82, 73734 Esslingen, Germany

M. Barkmeyer · A. Kaluza
Chair of Sustainable Manufacturing & Life Cycle Engineering, Institute of Machine Tools and Production Technology (IWF), Technische Universität Braunschweig, Langer Kamp 19b, 38106 Braunschweig, Germany

E. Knüpffer · S. Albrecht
Department of Life Cycle Engineering (GaBi), Fraunhofer Institute for Building Physics (IBP), Nobelstraße 12, 70563 Stuttgart, Germany

© Springer Nature Switzerland AG 2019
L. Schebek et al. (eds.), *Progress in Life Cycle Assessment*, Sustainable Production, Life Cycle Engineering and Management,
https://doi.org/10.1007/978-3-319-92237-9_14

the process assessment are categorized by a traffic light function, so that decision makers within the company will be able to integrate the results into their department and improve the processes they are responsible for.

Keywords Life cycle assessment · Social life cycle assessment · Life cycle working environment · Automation technology components

1 Introduction

More and more companies recognize the necessity to contribute to a sustainable development and integrate sustainability aspects into their core business. Thus, the consideration of the life cycle of products (upstream processes, production phase, use-phase, end-of-life phase) gains more relevance. However, most implemented methods focus on ecological and economic aspects. Social impacts along the value chain are likely to be disregarded due to methodological challenges with assessment of social impact (Kloepffer et al. 2009). The Life Cycle Working Environment (LCWE) solves this problem: it is based on statistical data and can be easily adapted to Life Cycle Assessment (LCA) studies because it builds on the results of its energy and material flows (Barthel 2014). As LCWE case studies only considered consumer goods so far (where no data of the use phase is available), the application is limited to upstream processes and the production phase (Albrecht et al. 2007). In contrast to consumer goods, business-to-business products and automation components (ATCs) respectively, have the property that they are also integrated into production process steps during use phase where data can be collected. Against this background, the first aim of this paper is to present how a social profile can be created for automation components in the three life cycle phases (upstream, production and use-phase) using the LCWE method. The second goal is to develop a procedure to integrate the LCWE method into a company context. Finally, it is presented to what extent the method can address internal strategic issues and how recommendations could be formulated for corporate purposes.

2 Life Cycle Working Environment Method

In general, Social Life Cycle Assessment (S-LCA) is a quantitative approach to assess social and socio-economic impacts of products along their life cycle (Albrecht et al. 2007; Benoit and Mazijn 2009; Jørgensen 2013). The Life Cycle Working Environment (LCWE) method developed by the Department of Life Cycle Engineering

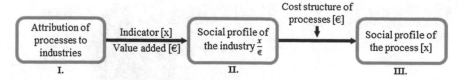

Fig. 1 Steps of LCWE method

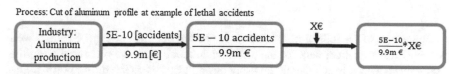

Fig. 2 Examplary application of the LCWE method

(GaBi) is one approach to perform S-LCA. LCWE studies result in social profiles, which describe one or more social indicator related to a specific process performed in a value-adding operation, e.g. within a company. One social indicator could be injuries that might occur during that operation (Barthel 2014). Statistical data serves as a foundation of the LCWE method. Currently, data from the statistical office Eurostat is applied. It is possible to complement the Eurostat data in LCWE by integrating further databases or primary data, e.g. from companies (Macombe et al. 2011).

Currently, LCWE is limited to six indicators but might be extended depending on data availability. The first indicator is the economic value added per process step. It is applied as an allocation key for the other indicators, which are (Albrecht et al. 2016):

- Lethal accidents
- Non-lethal accidents
- Working time
- Share of female workers
- Qualified working time

In general, some of the indicator values should be minimized e.g. accidents. Others depend on strategic considerations within the process setting, e.g. desired share of female workers within a company.

LCWE consists of three steps, as shown in Fig. 1. Figure 2 exemplarily shows the LCWE application to the cutting of an aluminum profile.

The first step of the LCWE an allocation of processes to industries is conducted. The considered database (Eurostat) provides data for various indicators separated by industries and topics. One topic is population and social conditions including labor related issues. This data serves as basis for the LCWE assessment. In the sample case, the cutting of an aluminium profile is attributed to the aluminium production industry (Barthel 2014).

Fig. 3 Examplary
pneumatic cylinder (Festo
2017)

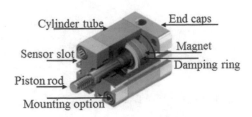

The values of the indicators (x) and values added per process step (€) are as well derived from statistical data for the different industries. In the presented case, the total value added through aluminum production in Europe is 9.9 million Euros. The LCWE indicators are exemplarily represented by the lethal accidents, showing a total value of 5E-10 accidents in Europe (Barthel 2014).

The indicators are then divided by the value added per industry to create social profiles of the industries (Barthel 2014).

Using the cost structure of the product and its functional unit respectively, the social profiles of the industries can be allocated to the social profiles of the processes within the respective product system. The cost structure of the LCWE represend the Life Cycle Inventory (LCI) of the LCA (Barthel 2014).

In the example, the cutting of an aluminum profile of one product causes costs of X Euros (Barthel 2014). By multiplying the share of cost of a single process with a social indicator per value added, the social profile of the process is generated (see Fig. 2) (Benoît and Mazijn 2009).

3 Application of LCWE for the Use Phase of Automation Technology Component

So far, LCWE has been used to calculate the social profiles of the upstream and production of product. This case study is the first one describing an integration of the use phase within LCWE.

3.1 Examplary Use Phase of Automation Technology Components

The paper is applied to assess a pneumatic cylinder as shown in Figs. 3 and 4.

There are many scenarios for the use phase of a cylinder, depending on its application in the manufacturing processes. The exemplary production system selected for this case study represents an assembly of a bottle top after filling (see Fig. 4). The use phase is separated in two phases, the use phase of the production system and the use phase of the operation phase of the system which is the assembly of a bottle top.

The system boundary covers the upstream processes, production and the described use phase. Due to the fact that the cylinder consists of nearly 70% aluminum, the aluminum waste is considered within the LCI. The functional unit is the production and use phase of one cylinder. The average lifetime is estimated with 5 million cycles which equals 5 million times assembling of bottle tops.

3.2 Procedure for the Application of the LCWE Method for Company Purposes

Procedure for the application of the LCWE method five parts are distinguished (see Fig. 5).

The **first part** of the procedure deals with the input of primary data. This includes available product information, e.g. the bill of material and the cost structure of the regarded product system. Moreover, it offers the company the opportunity to integrate primary data for single indicators, e.g. the number of fatal accidents. The input data formula gives the opportunity to integrate corporate goals for each indicator. Furthermore, the cost structure of the LCI is included as primary data. Only the costs related to the operating use phase are estimated.

Secondly a database, which includes data of the indicators and their cost structure, must be set up. The database therefore embraces the secondary statisti-

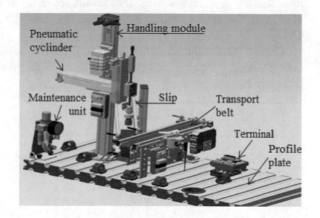

Fig. 4 Examplary production system for the use phase (Festo Didactic 2017)

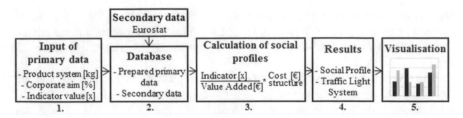

Fig. 5 Procedure for the application of the LCWE method for company purposes

cal data as well as the reprocessed data from the company from the first part. In this paper, secondary data is gathered from Eurostat (2017). This includes input data for the calculation of the indicators in upstream activities and the use phase.

The calculation of social profiles using the LCWE method is the **third part** of the procedure. This part defines the system boundaries of the procedure because the calculation works with data of the database and depends on the life cycle impact assessment and the system boundaries of each calculation. The LCI and the energy and material flows respectively are set up in a simplified manner to enable an accelerated use within the company. The integration of statistical data from Eurostat is also considered during LCI modeling, e.g. regarding the different industries involved. One limitation of the simplified LCI is that the LCI model does not account for recirculation of production scrap.

The **fourth part** represents the social profile of the processes as a result. The results are divided into the different life cycle phases and indicators. Moreover, in this part, the results of the calculation are compared with the corporate goals to get a classification within a traffic light system. A red traffic light signalizes that the results have a deviation higher than 10%, a yellow traffic light means that the results have a deviation between 1% and 10% and the green traffic light signalizes that the deviation is lower than 1%.

The **last part** of the procedure contains different visualizations of the results. The comparison between the social profiles and corporate goals regarding a social profile is depicted.

4   Visualisation and Integration into Company Processes

Figure 6 shows the qualitative results of the share of female workers. Furthermore, the results are set in comparison to exemplary corporate goals (assumptions for the current case). In this example, the production phase has the biggest share of women working time.

An objective weighting of social indicators is not possible For example, child labor compared to the number of lethal accidents evokes different opinions and uncertainties.

A possibility to handle the different results is to assign them to the different departments of a company. In order to assign the results to the right department there are two opportunities for visualization (see Table 1). Arrange the results based on the life cycle phases or arrange the results according to the target achievement. If the traffic light indicates red, a limit value is exceeded and there is a need for action. In case of a yellow traffic light, it is possible to decide whether an action should be carried out or not. The green traffic light represents the conformity with corporate goals. For example, the results of the red and yellow traffic lights are important for risk management because there might be a big risk for the company. The material management is interested in the yellow and red traffic lights up the upstream. The decisions markers within the company are now able to use this procedure as a decision support.

Fig. 6 LCWE result: Share of female workers

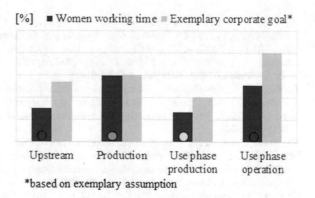

Table 1 Integration of LCWE results into company processes

Department	Upstream	Production	Use phase production	Use phase operation	Target achievement per indicator
Risk management	x	x	x	x	⦿̊
Product development & product management			x	x	⦿̊
Material management	x				⦿̊
Purchase & supplier management system	x				⦿̊
Marketing	x	x	x	x	⦿̊
Controlling	x	x	x	x	⦿̊
Sustainability department	x	x	x	x	⦿̊

5 Conclusion

Based on the inventory modeling of an environmental LCA, an S-LCA applying the LCWE method is performed. This paper is on the one hand about creating a concept for the integration of the LCWE for company purposes. On the other hand, recommendations on how to use the results within company processes are derived. A comparison between the results of the social profiles and the corporate goals can be conducted. The derived recommendations directly assign LCWE results to departments. It is designed for all kind of ATCs and allows the integration of new indicators. Moreover, the results of the concept could be helpful within strategic questions because they detect social hotspots and grievances along their value chain, which could be a threat to companies.

References

Albrecht, S., Barthel, L.-P., Baitz, M., Deimling, S., Fischer, M., Plieger, J.: The sustainability of packaging systems for fruit and vegetable transport in Europe based on life-cycle-analysis. InLCA/LCM, Portland, US (2007)

Albrecht, S., Endres, H.J., Knüpffer, E., Spierling, S.: Biokunststoffe–quo vadis?, in: Uwf—UmweltWirtschaftsForum, S.1–8 (2016)

Barthel, L-P.: Methode zur Abschätzung sozialer Aspekte in Lebenszyklusuntersuchungen auf Basis statistischer Daten, Dissertation, Fraunhofer Verlag, Stuttgart (2014)

Benoît, C., Mazijn, B. (eds.): Guidelines for social life cycle assessment of products. UNEP/SETAC, Belgien (2009)

Eurostat: Datenbank, Eurostat, http://ec.europa.eu/eurostat/de/data/database. (2017)

Festo Didactic.: Station Pick&Place–Klein und fein, http://www.festo-didactic.com/de-de/lernsysteme/mechatronische-systeme-mps/mps-stationen/station-pick-placeklein-und-fein.htm?fbid=ZGUuZGUuNTQ0LjEzLjE4LjYwNi44MTQ5MTQ5MTQ5MTQ5MTQ5MTQ1. (2007)

Festo: Kompaktzylinder ADN/AEN, ISO 21287, https://www.festo.com/cat/xdki/data/doc_de/PDF/DE/ADN_DE.PDF. (2017)

Jørgensen, A.: Social LCA—a way ahead? Int. J. Life Cycle Assess. **18**(2), 296–299 (2013). (Springer Verlag, Berlin/ Heidelberg)

Kloepffer, W., Grahl, O.: Ökobilanz (LCA). Ein Leitfaden für Ausbildung und Beruf, WILEY-VCH Verlag GmbH & Co. KAaA, Weinheim (2009)

Macombe, C., Feschet, P., Garrabé, M., Loeillet, D.: 2nd International seminar in social life cycle assessment—recent developments in assessing the social impacts of product life cycles. Int. J. Life Cycle Assess. **16**(9), 940–943 (2011). (Springer Verlag, Berlin/ Heidelberg)

Life Cycle Assessment of Industrial Cooling Towers

Christine Schulze, Sebastian Thiede and Christoph Herrmann

Abstract Sustainability and life cycle thinking is becoming part of companies' "green identity". For this reason, Life Cycle Assessment (LCA) is a common tool to assess the environmental impacts of manufactured products in order to improve it—starting with raw material selection to efficiency improvements of manufacturing processes and optimization of operation. Focusing the manufacturing stage, usually only the machines and processes directly related to the manufacturing are taken into account for assessment. Peripheral processes e.g. pump and pipe systems and technical building services (TBS) e.g. air conditioning, process water supply, are often neglected due to poor data availability. Supplying cooling water for production machines, cooling towers (CTs) are a central part of the industrial TBS—and therefore part of almost every manufacturing system. Neglecting the environmental impact of CT would result in a major white spot when accomplish a LCA. In order to close this gap, this paper presents the first LCA of industrial CTs showing the hot spots of environmental impacts along the entire CT life cycle.

Keywords Cooling tower · Energy demand · Fresh water demand

1 Introduction

Water is one of the most important elements in our daily life and basis of existence for almost every creature on our planet. The "blue planet" is named by the colour of the surface water, which covers 71% of the earth surface. The majority of existing water (97%) occurs as salted water in oceans. The remaining 3% are fresh water e.g. in lakes and rivers or locked up as arctic ice (Perlman 2016). Therefore, fresh water, useable for living purposes, is a very rare resource. Due to climate change, growing

C. Schulze (✉) · S. Thiede · C. Herrmann
Chair of Sustainable Manufacturing and Life Cycle Engineering, Institute of Machine
Tools and Production Technology (IWF), Technische Universität Braunschweig,
Langer Kamp 19b, 38106 Braunschweig, Germany
e-mail: ch.schulze@tu-braunschweig.de

© Springer Nature Switzerland AG 2019
L. Schebek et al. (eds.), *Progress in Life Cycle Assessment*, Sustainable
Production, Life Cycle Engineering and Management,
https://doi.org/10.1007/978-3-319-92237-9_15

135

population and industrialization, water stress in the year of 2040 is predicted by the World Resource Institute (Luo et al. 2015) for majorly industrialized countries, e.g. USA, China, Australia and European countries such as Spain and Italy. Therefore, the use of water is more and more in focus when companies develop their factory related sustainability strategies (Dehning et al. 2016; Kurle et al. 2015).

In industrial context, water is widely used as operational resource, e.g. for washing processes and heat transfer. The usage of water is one main drivers for total energy consumption in industrial process chains (Thiede et al. 2016). The so-called water-energy-nexus describes the relation of water and energy: energy for water, such as water treatment, pumping, heating etc., water for energy, which is used for e.g. extraction, purification, processing of natural gas, coal and electricity generation (Thiede et al. 2017; Walsh et al. 2015). For LCA in production environment, water is usually only considered as resource for the product with rather neutral impact in case of energy flows. Although, the main share of water is not directly used in products, but for energy and mass transfer in the discrete manufacturing (Kurle et al. 2017). Thus, it is necessary to consider water and water-energy-nexus related devices in factories for LCA studies as well as several studies demonstrated (Lévová and Hauschild 2011; Mousavi et al. 2015). Industrial cooling towers (CT) are an example for the water-energy nexus in manufacturing. Several studies have researched Key Performance Indicators (KPI) for operation and control during the usage stage of CT (Thiede et al. 2017; Cortinovis et al. 2009; Schlei-Peters et al. 2017; Guo et al. 2017; Kurle 2018). The studies prove high environmental impacts in green house gas emissions as well as fresh water demand depending on the operation mode as well as the climatic condition of the location (Schulze et al. 2018). Nevertheless, for a holistic assessment of the environmental impacts of CTs all life cycle stages should be taken into account. For that reason, this study presents an LCA for industrial CTs.

2   Background

2.1   Industrial Cooling Tower Systems

In industrial plants, cooling is demanded by many places. These include in particular machines that require cooling for their components or processes. Since these machines are distributed throughout the entire plant location, often decentralized cooling-supply is required. As a rule, CT systems are used if the waste heat is not available for further use (Kurle et al. 2016). A detailed overview of the components, energy and mass flows can be found in Fig. 1. The heated water from the production machines is supplied via pump and pipe system and sprayed into the upper part of the CT. In counter flow, ambient air is forced-draft in by a fan. Through energy and mass transfer between water and air, the water is cooled down while the air is saturated with evaporating water. In order to maximize the time for energy and mass transfer, fillers are installed in the CT where the water slowly flows down. Finally, the cooled

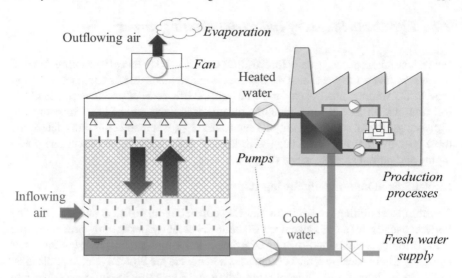

Fig. 1 Elements of industrial CTs (Schulze et al. 2018)

water is pumped back to the production machines. As thermodynamically open systems, the operation of CT is highly impacted by the environmental conditions of the location. Warm and humid climate impairs the energy and mass transfer leading to higher air demand and fan operation (Schulze et al. 2018).

For industrial purposes, different types of re-cooling systems are broadly differentiated by the type of coolant: dry or wet as well as the type of air supply natural-draft ventilation or forced-draft ventilation. Hence, different construction types are developed for the special field of application. Dry coolers, using only the ambient air as coolant, are used for low cooling requirements or flexible locations without large-scale infrastructure use. Thereby ventilators are used to ventilate the objects to be cooled, e.g. for slow cooling of products. Due to the low heat capacity of air, the cooling capacity of dry CT is limited (Schodorf and Geiger 2012). If larger waste heat streams need to be cooled, coolants with higher heat capacity are used. As a technical medium with one of the highest heat capacities, fresh water is often used as a coolant. In the so-called wet CT types, the warm water is cooled in the ambient air while partly evaporate the water. The evaporation increases the cooling capacity and therefore wet CT can be used also for higher ambient heat temperatures (24–26 °C) during the summer time (Schodorf and Geiger 2012). Furthermore, a particular distinction is made between the types of air supply: natural-draft by using the so-called chimney effect and force-draft ventilation by fan. CTs with forced-draft ventilation are usually used at industrial sites as described above. In contrast to other types, the design can be significantly more compact and requires less space at factories' sites. In addition, the formation of visible vapour can be avoided. Vapours can be carriers of legionellae, which may be a danger for humans in the workplace.

2.2 Life Cycle Stages of Industrial Cooling Tower

For industrial purposes, usually forced-draft CT are used to meet the cooling demand of lower temperature waste heat from production machines. In this compact-designed type of CT water is used as coolant. The generic life cycle of an industrial used CT has four stages: raw material, manufacturing, usage and end of life. For every life cycle stage several inputs and outputs such as energy, water and emissions determine the environmental impacts related with the CT system. In the next sections, these inputs and outputs are described (Fig. 2).

Raw material and manufacturing stage

The compact building type of industrial CTs consist mainly of steel and plastics. The body is usually made of galvanized or stainless steel by performing processes like cold forming, welding and riveting. Further installations such as nozzles for water spraying, fillers packages, mist eliminators and covers are typically made of plastics, mainly polyvinylchloride. Injection moulding and film extrusion processes are main processes used for their production. The water supply is realized by a pump and pipe system towards and back from the production machines. For pump and fan engines, electrical motors are used (Schodorf and Geiger 2012; Berliner 1975). A scheme of a typical industrial CT with its main parts is presented in Fig. 1.

Usage stage

The usage stage of a CT can last some decades. Usually TBSs are operated permanently during production time of the manufacturing processes, which can be for instance six days with three shifts a week. During the operation time, the coolant water is pumped to the top of the CT and air flow is ventilated in counter flow direction by the fan. For the operation of pumps and fan, electrical energy is consumed. In addition, due to evaporation and blow-down the fresh water demand of CT is a remarkably factor for environmental impacts. Typical industrial CT operate with water as coolant, which is taken either from groundwater or from public water supply.

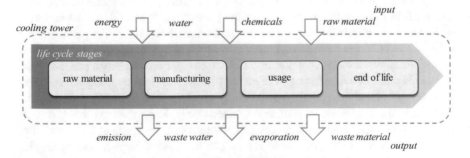

Fig. 2 Generic life cycle stages of industrial CTs, inputs and outputs

Contaminated cooling water is removed as waste water by blow-down and replaced with fresh water. The fillers are replaced regularly by maintenance.

End of life stage

At the end of the usage stage, the CT can completely disassembled. The steel body can be treated as scrap steel. The electric motors of the pumps and fan can be treated as electric scrap. The fillers are treated as scrap plastic.

2.3   Approaches for Life Cycle Assessments of Industrial Cooling Towers

Different approaches and studies looking at environmental impacts from CT have been identified. Several authors assess the usage of CTs as part of a building or energy generation system. Many approaches particularly focus on water losses through evaporation as well as resulting fresh water demand (Burkhardt et al. 2011; Koroneos and Tsarouhis 2012; Guerra et al. 2014; Ali and Kumar 2016). In industrial context, several studies have been published focusing on the energy and water efficiency of industrial CTs during the use stage (Thiede et al. 2017; Kurle et al. 2017; Schlei-Peters et al. 2017). In Schodorf and Geiger (2012) the usage stage of different types of CT are compared regarding energy demand, CO_2 footprint and costs to give a broad overview of the pro's and con's of each type. For the assessment of the economic efficiency and CO2 footprint of evaporation and hybrid CT, the Mechanical Engineering Industry Association (VDMA) published a guideline (Markus et al. 2015). It also focuses on the use stage of CT, arguing that CO_2 emissions from manufacturing stage would account less than 1% of the total CO_2 emissions over the whole life cycle. Schulze et al. present an analysis of the environmental impacts from industrial CTs operation considering individual climate data and energy mixes for different locations (Schulze et al. 2018). Nevertheless, for the holistic assessment of environmental impacts over the whole lifetime a life cycle assessment is necessary.

3   Life Cycle Assessment of Industrial Cooling Towers

In order to perform an LCA study, the regarded CT system with its system boundaries is described and goal and scope are defined. Subsequently, the underlying data for the Life Cycle Inventory (LCI) is presented followed by the results of the Life Cycle Impact Assessment (LCIA). The model for LCIA was built in the software environment UMBERTO.

3.1   Goal and Scope Definition

The object of this case study is an industrial force-draft CT as part of the TBS of an automotive production plant located in Germany. The CT system includes pump and pipe system, transporting the warm water to and back from the production. Furthermore, the air-draft is soken by the fan located on the top of the cooling tower in counter flow direction to the water. In order to increase the time for energy and mass transfer, the water rinses down over filler built into the cooling tower. As the outcome of CT operation is cooled water, the cooling of 1 kg water from 35 to 28 °C in Germany for the overall usage time is chosen as the functional unit for the LCA. The impact assessment has been conducted using CML 2001 methodology.

As it is assumed that all processes of the life cycle take place in Germany, data for this area have been used. For raw materials and manufacturing of industrial CT, mainly stainless steel coils for the body and polyvinylchloride films (PVC) for the filler packages are considered. The fillers are replaced completely every 5 years as maintenance and prevention for clogging. For the peripheral aggregates, electric motors for pumps and a fan are taken into account. Due to the individual CT design, the pipe system is neglected in this use case. The usage time of this CT accounts for 20 years. During this time, the electricity demand of pumps and fan are considered as energy input flows. Furthermore, regular exchange of cooling water causes fresh water demand as well as waste water generation. As in this case study the CT is located in Germany, data for the energy and fresh water demand for the location of Berlin are taken from Schulze et al. (2018). In end-of-life, a treatment of the scrap is considered for the fractions steel, plastics and electrical components. The details and data for LCI are listed in Table 1.

3.2   Life Cycle Inventory Analysis of Industrial Cooling Towers

In order to analyse the LCI, the flows of energy and material in each life cycle stage have to be determined. Adequate datasets have been selected from Ecoinvent 3 database for modelling the resource demand in UMBERTO (Ecoinvent 2016).

3.3   Results of the Life Cycle Impact Assessment (LCIA)

The impact assessment has been conducted using CML 2001 methodology. The LCIA results are presented exemplarily in Fig. 3, as well as detailed in Tables 2, 3, 4 and 5 for the categories climate change GWP 100a, human toxicity HTP 100a, freshwater aquatic ecotoxicity, FAETP 100a, terrestrial ecotoxicity TAETP 100a.

The results coincide that the environmental impacts primarily occur during the usage stage of the industrial CT. The life cycle stages of raw material and manufacturing as well as end of life only play minor roles here. These findings correlate with

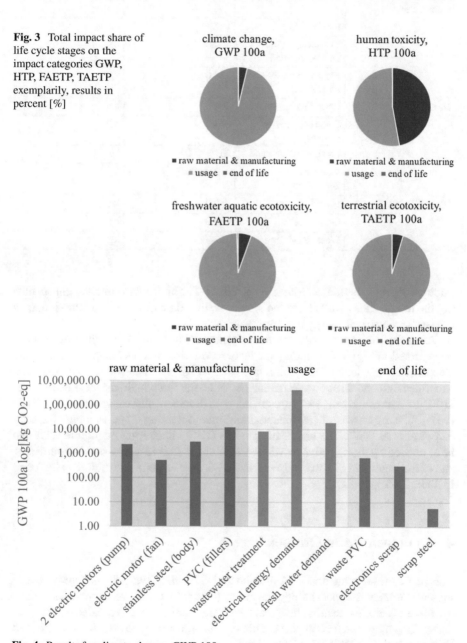

Fig. 3 Total impact share of life cycle stages on the impact categories GWP, HTP, FAETP, TAETP exemplarily, results in percent [%]

Fig. 4 Results for climate change, GWP 100a

Table 1 LCI of the industrial CT (Schulze et al. 2018)

Life cycle stage	Objective	Amount	Unit
Raw materials and manufacturing	Stainless steel (body)	850.0	kg
	PVC (fillers)	192.3	kg
	2 electrical motors 18 kW (pumps)	600.0	kg
	Electrical motor 15 kW (fan)	135.0	kg
Usage	Operating hours	8500	h/a
	Electrical energy demand	864.0	MWh
	Fresh water demand	63740000.0	kg
	Waste water generation	21246000.7	kg
End of life	Electric scrap	735.0	kg
	Scrap steel	850.0	kg
	Waste PVC	192.3	kg

findings of previous studies (Markus et al. 2015). The high environmental impact during the usage stage results from the long operation time (20 years and continuously demand of electrical energy and fresh water).

Focussing the impact category climate change, GWP 100a, a contribution analysis is presented in Fig. 4. Comparing the resource in- and outputs listed in Table 1, the electrical energy demand is by far related to the highest GWP, followed by the fresh water demand and waste water treatment. Thus, the three main driver processes for GWP are part of the usage stage, explaining its dominant share regarding total GWP (96%). The processes of raw material and manufacturing stage accounts for only 3.75% of total GWP. The end of life process GWP impacts are negligible. This can be explained by the effective recycling treatment of stainless steel and low-emission end of life treatment of PVC and electrical devices. The detailed results can be found in Table 2 (Appendix).

4 Conclusion and Outlook

This paper presents the first LCA of industrial CTs, identifying the hot spots regarding environmental impacts along the entire life cycle. Confirming previous studies reviewed before, the results show that the usage stage has been revealed as the main driver for environmental impacts. This conclusion can be drawn for the selected impact categories GWP, HTP, FAETP and TAETP respectively. This can be explained by the long usage time of 20 years, which is not unusual for TBS devices. With a

Table 2 LCIA results for climate change, GWP 100a [kg CO_2-eq]

Life cycle stage	Processes of CT life cycle	[kg CO_2-eq]	Total share (%)
Raw material and manufacturing	2 electric motors (pump)	2350.92	3.75
	Electric motor (fan)	528.96	
	Stainless steel (body)	3036.88	
	PVC (fillers)	11,821.11	
Usage	Wastewater treatment	8157.32	96.04
	Electrical energy demand	427,639.13	
	Fresh water demand	18,766.55	
End of life	Waste PVC	682.85	0.14
	Electronics scrap	314.01	
	Scrap steel	5.73	
Results		473,303.46	100.00

zoom in for GWP, the electrical energy demand for pump and fan operation is the main driver of environmental impact by far. Furthermore, water-based objectives as fresh water demand and water treatment are at the next positions for environmental impact. As mentioned before, cooling tower is an example of the water-energy nexus in production, it is necessary to consider water in LCA studies. As CT operation is highly impacted by the environmental conditions, different locations should be considered for future LCA work as well. As the environmental footprint of countries' specific electricity generation mix is typically in continuous change, in particular due to a significant increase in installations of renewable energies, further work should consider future electricity mixes as well. A first study considering different locations and energy mixes is given by Schulze et al. (2018), focussing the usage stage only. Furthermore, there are various types of CT which entail different environmental impacts (Schodorf and Geiger 2012). This study has been applied for an open circuit draft-forced CT only, which is one of the main applied types for industrial purposes. For further studies, the authors propose the extension of the objective to a broaden range of CT types.

Appendix

See Tables 3, 4 and 5.

Table 3 LCIA results for freshwater aquatic ecotoxicity, FAETP 100a [kg 1.4-DCB-Eq]

Life cycle stage	Processes of CT life cycle	[kg 1.4-DCB-Eq]	Total share (%)
Raw material and manufacturing	2 electric motors (pump)	4714.00	5.62
	Electric motor (fan)	1060.65	
	Stainless steel (body)	8904.99	
	PVC (fillers)	4166.97	
Usage	Wastewater treatment	6647.30	93.81
	Electrical energy demand	296,771.86	
	Fresh water demand	10,987.25	
End of life	Waste PVC	813.65	0.24
	Electronics scrap	1019.22	
	Scrap steel	77.47	
Results		335,163.35	100.00

Table 4 LCIA results for human toxicity, HTP 100a [kg 1.4-DCB-Eq]

Life cycle stage	Processes of CT life cycle	[kg 1.4-DCB-Eq]	Total share (%)
Raw material and manufacturing	2 electric motors (pump)	12,094.76809	48.51
	Electric motor (fan)	2721.32282	
	Stainless steel (body)	48,993.06576	
	PVC (fillers)	7577.173661	
Usage	Wastewater treatment	11,474.86226	51.37
	Electrical energy demand	56,542.81797	
	Fresh water demand	7577.173661	
End of life	Waste PVC	63.67050381	0.04
	Electronics scrap	117.8246506	
	Scrap steel	2.872899321	
Results		147,165.55	100.00

Table 5 LCIA results for terrestrial ecotoxicity, TAETP 100a [kg 1.4-DCB-Eq]

Life cycle stage	Processes of CT life cycle	[kg 1.4-DCB-Eq]	Total share (%)
Raw material and manufacturing	2 electric motors (pump)	4.432153852	4.68
	Electric motor (fan)	0.997234617	
	Stainless steel (body)	2.664751244	
	PVC (fillers)	2.41021486	
Usage	Wastewater treatment	21.26068213	95.23
	Electrical energy demand	169.5303375	
	Fresh water demand	23.10492157	
End of life	Waste PVC	0.181955083	0.08
	Electronics scrap	0.032152093	
	Scrap steel	0.005028801	
Results		224.62	100.00

References

Ali, B., Kumar, A.: Development of life cycle water footprints for gas-fired power generation technologies. Energy Convers. Manag. **110** (2016)

Berliner, P.: Kühltürme, Grundlagen der Berechnung und Konstruktion. Springer, Karlsruhe (1975)

Burkhardt, J.J., Heath, G.A., Turchi, C.S.: Life cycle assessment of a parabolic trough concentrating solar power plant and the impacts of key design alternatives. Environ. Sci. Technol. **45**(6), 2457–2464 (2011)

Cortinovis, G.F., Paiva, J.L., Song, T.W., Pinto, J.M.: A systemic approach for optimal cooling tower operation. Energy Convers. Manag. **50**(9), 2200–2209 (2009)

Dehning, P., Lubinetzki, K., Thiede, S., Herrmann, C.: Achieving environmental performance goals—Evaluation of impact factors using a knowledge discovery in databases approach. Procedia CIRP **48**, 230–235 (2016)

Ecoinvent: Ecoinvent 3.3 dataset documentation (2016) [Online]. Available: http://ecoinvent.ch/. Accessed 20 Sept 2017

Guerra, J.P.M., Coleta, J.R., Arruda, L.C.M., Silva, G.A., Kulay, L.: Comparative analysis of electricity cogeneration scenarios in sugarcane production by LCA. Int. J. Life Cycle Assess. 1–12 (2014)

Guo, Y., Wang, F., Jia, M., Zhang, S.: Parallel hybrid model for mechanical draft counter flow wet-cooling tower. Appl. Therm. Eng. **125**, 1379–1388 (2017)

Koroneos, C., Tsarouhis, M.: Exergy analysis and life cycle assessment of solar heating and cooling systems in the building environment. J. Clean. Prod. **32**, 52–60 (2012)

Kurle, D.: Integrated Planning of Heat Flows in Production Systems. Springer International Publishing (2018)

Kurle, D., Thiede, S., Herrmann, C.: A tool-supported approach towards water efficiency in manufacturing. Procedia CIRP **28**, 34–39 (2015)

Kurle, D., Schulze, C., Herrmann, C., Thiede, S.: Unlocking waste heat potentials in manufacturing. Procedia CIRP **48**, 289–294 (2016)

Kurle, D., Herrmann, C., Thiede, S.: Unlocking water efficiency improvements in manufacturing—From approach to tool support. CIRP J. Manuf. Sci. Technol. (2017)

Lévová, T., Hauschild, M.Z.: Assessing the impacts of industrial water use in life cycle assessment. CIRP Ann. - Manuf. Technol. **60**(1), 29–32 (2011)

Luo, T., Young, R., Reig, P.: Aqueduct projected water stress country rankings. World Resour. Inst., no. August, pp. 1–16 (2015)

Markus, N., et al.: Economic analysis and partial carbon footprint of evaporative cooling equipment—Guideline for calculation (2015)

Mousavi, S., Kara, S., Kornfeld, B.: Assessing the impact of embodied water in manufacturing systems. Procedia CIRP **29**, 80–85 (2015)

Perlman, H.: Where is Earth's water? USGS Water-Science School. U.S. Department of the Interior, U.S. Geological Survey (2016) [Online]. Available: http://ga.water.usgs.gov/edu/earth wherewater.html. Accessed 12 Feb 2018

Schlei-Peters, I., Wichmann, M.G., Matthes, I.-G., Gundlach, F.-W., Spengler, T.S.: Integrated material flow analysis and process modeling to increase energy and water efficiency of industrial cooling water systems. J. Ind. Ecol. **00**, 1–14 (2017)

Schodorf, W., Geiger, K.: Kühl- und Kältetechnik – investieren um zu sparen. IHKS Fach.Journal Klima/Kälte 46–51 (2012)

Schulze, C., Raabe, B., Herrmann, C., Thiede, S.: Environmental impacts of cooling tower operations—The influence of regional conditions on energy and water demands. In: Procedia CIRP 25th Conference on Life Cycle Engineering, vol. 69, pp. 277–282 (2018)

Thiede, S., Schönemann, M., Kurle, D., Herrmann, C.: Multi-level simulation in manufacturing companies: the water-energy nexus case. J. Clean. Prod. **139** (2016)

Thiede, S., Kurle, D., Herrmann, C.: The water–energy nexus in manufacturing systems: framework and systematic improvement approach. CIRP Ann. - Manuf. Technol. (2017)

Walsh, B.P., Murray, S.N., O'Sullivan, D.T.J.: The water energy nexus, an ISO50001 water case study and the need for a water value system. Water Resour. Ind. **10**, 15–28 (2015)

Bioplastics and Circular Economy—Performance Indicators to Identify Optimal Pathways

Sebastian Spierling, Venkateshwaran Venkatachalam, Hannah Behnsen, Christoph Herrmann and Hans-Josef Endres

Abstract With a growing demand of resources and environmental issues like climate change, circular economy has become an inevitable fundamental principle in resource consumption. While conventional plastics have become an essential material group in the 20th century, bio-based and/or biodegradable alternatives are still evolving. Principles of the circular economy need to be implemented for all kinds of material flows and bioplastic is no exception, as biomass is a limited resource as well. To identifiy environmentally favourable pathways for bioplastic waste streams, it is helpful to develop suitable specific indicators. To review the current status quo, a literature review was conducted with focus on (bio-) plastics. While general circular economy approaches are available manifold, only few literature on specific circular economy indicators for (bio-) plastics is available. Especially the aspect of biodegradability is unique for bioplastics and therefore enables further waste treatment options. An already existing framework is therefore extended for these waste treatment options to enable the identification of optimal pathways also for biodegradable plastics.

Keywords LCA · Bioplastics · Circular economy

1 Introduction

Materials enable us to meet our daily demands like housing, transport, communication or packaging. With a growing world population as well as economic development, the demand for materials is predicted to more than double from 2015 to 2050 (UNEP 2017). A material group, which has become more and more essential

S. Spierling (✉) · V. Venkatachalam · H. Behnsen · H.-J. Endres
Institute for Bioplastics and Biocomposites, University of Applied Sciences and Arts Hannover, Heisterbergallee 10a, 30453 Hannover, Germany
e-mail: Sebastian.spierling@hs-hannover.de

C. Herrmann
Institute of Machine Tools and Production Technology, Chair of Sustainable
Manufacturing & Life Cycle Engineering, Technische Universität Braunschweig, Langer Kamp 19 B, 38106 Braunschweig, Germany

© Springer Nature Switzerland AG 2019
L. Schebek et al. (eds.), *Progress in Life Cycle Assessment*, Sustainable
Production, Life Cycle Engineering and Management,
https://doi.org/10.1007/978-3-319-92237-9_16

in the 20th century, are plastics (Geyer et al. 2017). In 2015, the global production of plastics has been 322 million tonnes and is estimated to further grow within the next years. Plastics are versatile materials, which are used in many different application areas like packaging, building and construction, automotive, medical or electrical & electronic appliances. In Europe, the market is dominated by short life applications like packaging with 39.9% of total plastic material demand (PlasticsEurope 2016). Currently the main share of these plastics is made up by fossil-based resources like crude oil. Around 4% of the world's annual oil production is used for plastics (Kreiger et al. 2014). Against the background of the finiteness of fossil resources as well as the challenges of global warming, the development of bio-based alternatives has been in the focus of research in recent years (BMBF 2014).

Bioplastics can be either bio-based and biodegradable (e.g. PLA—Polylactide) or bio-based and durable (e.g. Bio-PE—Bio-Polyetyhlene). The third type of bioplastic is fossil-based and biodegradable, e.g. PCL (Polycaprolactone) (Endres and Siebert-Raths 2011). Furthermore, it is important to distinguish between chemical novel bioplastics and so called drop-in bioplastics. Chemical novel bioplastics, (e.g. PHA—polyhydroxyalkanoate) have a unique chemical structure and therefore have unique properties, processing profiles as well as waste management demands. The drop-in types (e.g. Bio-PE) have the same chemical structure as their conventional counterparts and differ only in terms of feedstock's (biomass instead of fossil resources). Therefore, their processing, utilization and disposal properties are identical to the conventional ones. In 2016, the production capacity of bioplastics amounts to merely 2.05 million tonnes and therefore constitutes currently just a small market share in comparison with conventional plastics. A growth of up to 9.20 million tonnes of bioplastics has been forecasted for 2021. The main share of bioplastics are bio-based plastics, which are based on biomass feedstock like sugar cane or corn (IfBB 2017). With a further evolving bioeconomy, the overall demand for biomass will increase as its usage in fuels, energy, heating and materials will broaden (Scarlat et al. 2015). However, given the limited biomass production capacities, resource efficiency and circular economy thinking are not only immanent for limited fossil-based product systems but also for bioeconomy ones like bioplastics. Defining what a 'Circular Economy' (CE) is, needs a multi-faceted approach. There exist different definitions for CE and there have been instances, where the understanding of CE (Geissdoerfer et al. 2017) and its different definitions have been extensively analyzed (Kirchherr et al. 2017). From these publications, it can be found out that defining a CE can be subjective and there are various possibilities for defining CE (Lieder and Rashid 2016). Out of all these definitions, the most famous definition of CE has been provided by Ellen MacArthur Foundation (Ellen MacArthur 2012), which reads: "CE is an industrial system that is restorative or regenerative by intention and design. It replaces the 'end-of-life' concept with restoration, shifts towards the use of renewable energy, eliminates the use of toxic chemicals, which impair reuse, and aims for the elimination of waste through the superior design of materials, products, systems, and, within this, business models". While plastic production capacities are growing, the circular economy thinking has not been fully implemented yet (World Economic Forum 2016). This is also applicable for bioplastics. As highlighted by

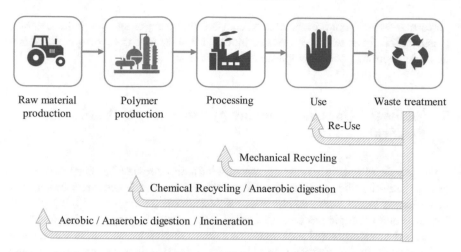

Fig. 1 Life cycle of bioplastics and loop of different waste treatment options

Geyer et al. (2017) in the timeframe of 1950 until 2015, about 8300 million tonnes of plastic have been produced. The main share of these plastics (4900 million tonnes) has been discarded (e.g. in landfill or litter into the environment), while 800 million tonnes were incinerated and just 600 million tonnes recycled. Another 2500 million tonnes are still in stock.

Therefore, it is indispensable to increase circular economy thinking within the plastic economy. While this is partly being implemented for the conventional plastics, bioplastics should also be involved. While the drop-in types can highly benefit from developments in the conventional plastic sector with regards to increase of recycling technologies etc., the chemical novel types have to be kept in the research focus. Furthermore, bioplastics offer additional waste treatment options in the form of aerobic and anaerobic digestion. In Fig. 1 it is highlighted on which stage of the value chain, different waste treatment options can be fed back in the bioplastic life cycle.

It is also important to distinguish between pre- and post-consumer (bio-) plastic waste. Pre-consumer waste occurs in the polymer production and processing step (e.g. off-grade material or scrap) and is usually not subject to sorting as it is homogeneous (same plastic type). Post-consumer waste occurs after the use phase and can be contaminated e.g. with food waste, is subject to collection and sorting and therefore has usually a low homogeneity.

To develop optimal pathways for bioplastics within a circular economy, it is important to develop indicators, which take technical implications on material properties as well as environmental or sustainable (environmental, social and economic) impacts of different waste treatment options into account. This study aims to contribute to a stronger integration of circular economy thinking into the bioplastic development and thereby creating more sustainable bioplastic value chains. To do so, the study will review current approaches to identify optimal circular economy pathways (e.g.

performance indicators) in the area of conventional plastics and adapt them to bio-plastics. The focus will be set on the biodegradability as a unique aspect of waste treatment options of some bioplastic types.

2 Review of Circular Economy Approaches for (Bio-) Plastics

To review the circular economy approaches, a literature review has been done using established scientific search engines like web of science, science direct or springer link using a search word combination of 'circular economy' and '(bio-) plastic' terms. While the topic has been discussed theoretically in many studies, few have developed or incorporated indicators. To assess the circularity, resource efficiency and environmental performance of bioplastics, it is essential to develop metrics and performance indicators for the corresponding products. Several approaches to assess the circularity of a product system have been developed so far. The cyclical use rate indicator for the reused and recycled material input (Ministry of the Environment, Japan 2003), the material circularity indicator to assess circularity at product and organization levels (Ellen MacArthur Foundation and Granta 2015) and the circular economy index to measure circularity based on the recycled material value (Di Maio and Rem 2015), to name a few. However, there is no standardized method for measuring the circularity of products (Linder et al. 2017) as of now. In addition, most of the above-mentioned studies are generic and are not product specific, which makes it difficult to interpret and apply them to plastics.

With respect to (bio-) plastics, only two recent studies were identified, that also applied indicators: Huysman et al. (2017) and Hildebrandt et al. (2017). Therefore, these will be discussed in detail and developed further. Huysman et al. (2017) have developed a so-called circular economy performance indicator, which is applied to a case study for post-industrial (pre-consumer) plastic waste. The proposed circular economy performance indicator (CPI) is defined as quotient of actual environmental benefit of the current applied waste treatment option and the ideal environmental benefit according to quality. The actual environmental benefit differs from the ideal one by taking recycling rates, quality loss as well as environmental impact of the recycling process into account. Following the ISO 14044 standard's (ISO 2006) distinction in closed-loop and open-loop recycling, the waste treatment options are classified into 4 options (also refer to Table 1: Options Ib, II, III and IV). Which option is most suitable for a plastic depends on the technical quality of the plastic. The technical quality is solely based on the aspect of miscibility of plastics and plastic blends. As the focus is set on conventional plastics, the additional waste treatment options enabled by biodegradability are not taken into account. While miscibility is an important aspect, other technical properties should also be taken into account and might differ in importance depending on the product applications. Furthermore, the recycling processes are mainly focusing on mechanical recycling and novel options

Table 1 Extended framework for circular economy and bioplastics (adapted from Huysman et al. (2017) for bioplastics)

Option	Waste treatment option	Actual environmental benefit	Ideal environmental benefit
Option Ia*	Re-use	$V_\alpha - R$	V_α
Option Ib	Mechanical recycling	$r \times V_\alpha - R$	V_α
Option Ic*	Chemical recycling, Anaerobic digestion	$r \times V_\alpha - R - T$	V_α
Option II	Mechanical recycling	$r \times p \times V_\alpha - R$	$p \times V_\alpha$
Option III	Mechanical recycling	$r \times m \times V_\beta - R$	$m \times V_\beta$
Option IV	Incineration, Anaerobic digestion	$E - I$	E
Option V*	Aerobic digestion	$B - I$	B

B Benefit for produced biomass; E Obtained amount of energy (Electricity and Heat); I Impact of incineration/anerobic and aerobic digestion process; m Mass balance to material substituted; p Maximal percentage of substituted virgin material; r Recycling rate; R Environmental impact of recycling proceess/refurbishment; T Treatment impact back to plastic; V_α Avoided impact of virgin material; V_β Avoided impact of secondary material
*Extension for the circular economy performance indicator framework of Huysman et al. (2017)

like chemical recycling should also be considered. On pre-consumer level, companies like NatureWorks conduct chemical recycling for off grade PLA on an industrial scale (NatureWorks 2017). Hildebrandt et al. (2017) focus on cascade use rather than circular economy. However, the focus of the work is set on bioplastics. The study highlights a novel procedure for calculating performance metrics along the cascade use chain for polymers. The proposed indicators include the cumulative energy demand, rate of material use efficiency, rate of energy recovery and substitution of virgin materials. While indicators are set up, they are not applied on the different waste treatment options for bioplastics or a case study.

3 Extended Framework for Circular Economy and Bioplastics

Bioplastics can offer additional waste treatment options in comparison to conventional plastics. While drop-in bioplastics can use the existing waste treatment structures of conventional plastics the chemical novel plastics might need modified structures or additional collection streams and separation processes. Biodegradable plastics can undergo aerobic (industrial/home composting) or anaerobic digestion (biogas plant) and thus provide further options of waste stream management. Therefore, it is important to develop circular economy performance indicators, which also include the biodegradable types of bioplastics. In Table 1, an advanced framework for covering waste treatment options of biodegradable plastics extended from Huysman et al. (2017) is presented. The extensions are highlighted with a "*".

The framework from Huysman et al. (2017) has been extended for the waste treatment options aerobic and anaerobic digestion, which are unique to bioplastics. Furthermore, chemical recycling and re-use were added, as it is important to consider them in a circular economy. However, these waste treatment options are also suitable for conventional plastics. In extension to the framework, options Ia, Ic and V have been added. Following the Huysman et al. (2017) approach, the quotient of actual environmental benefit and ideal environmental benefit form the circular economy performance. An indicator with "1" as value is therefore an ideal circular economy option. The new proposed framework is set up for pre-consumer recycling. Post-consumer recycling has to take collection and sorting into account, which might differ for various waste treatment options. In addition, it is important to highlight that not all bioplastics are suitable for all pathways (e.g. Option Ib—anaerobic digestion is suitable for a few bioplastics like PHA but not for durable biobased plastics like Bio-PE, as PHA is biodegradable and resulting biogas can be fed back into the PHA value chain). For option Ia, the waste treatment option is re-use. The actual benefit is set up of Vα (avoided impact of virgin material) minus the environmental impact of the refurbishment process (R). Accordingly, the ideal environmental benefit is Vα. Option Ib is mechanical recycling, taking the recycling rate as well as the environmental impact of the recycling process into account. Option Ic for chemical recycling and anaerobic digestion also takes the impacts of the processing back to plastic grade into account as both waste treatment options generate intermediates like biogas or different bio-based oligo-or monomors, which can be used as feedstock for a new (re-)polymerization process. However, the resulting material has therefore the same properties as virgin material and no quality loss. Option II and option III consider mechanical recycling of lower quality, where either additional virgin material is needed or another secondary material is substituted. Option IV include incineration and anaerobic digestion. The actual environmental benefit is calculated with obtained amount of energy (electricity and heat) from direct incineration of plastic or biogas minus impact of incineration or biogas process. This means that option IV focuses on the energy usage of the reaction products of the anaerobic digestion process, while option Ic focuses on their usability as polymer feedstock. Option V is aerobic digestion with the actual benefit of the benefit from produced biomass as a digestion residue minus impact of aerobic digestion process (composting). The potential benefit and associated credit of the created biomass during aerobic digestion have to be further investigated. However potential benefits of biodegradability e.g. in connection with marine litter are not taken into account in this framework.

The highlighted formulas can be used to calculate the CPI for different waste treatment options and impact categories of bioplastics. Thereby more optimal pathways from technical and environmental perspective can be chosen. A challenge however is the incorporation of technical aspects. Huysman et al. (2017) use the quality factor to assess which option is, from a technical point of view, best suitable for the waste stream. However, Huysman's proposed quality factor is solely based on compatibility (with current assumptions like binary compounds), just preliminary and needs further research. Beside other technical quality aspects, it is also important to assess the impact of a waste treatment option on the next waste treatment phase in further life cycles stages (cascade use). As different waste treatment options might have a different impact on this issue, such aspects also have to be considered. E.g. mechanical recycling of (bio-) plastics might cause the reduction of material quality in several life cycle stages, while chemical recycling has the opportunity to create recycled material without significant quality loss.

4 Conclusion and Outlook

To identify optimal circular economy pathways, it is important to develop performance indicators. For bioplastics, no performance indicators have been developed yet. As bioplastic value chains are currently established, it is important to provide this information to foster circular economy. As bioplastics provide additional waste treatment options like biodegradation, indicators have to be included also for aerobic and anaerobic digestion. The proposed framework provides indicators for bioplastics and considers these aspects for pre-consumer waste. Seven formulas enable to determine the circular performance for all available waste treatment options for bioplastics. When considering post-consumer (bio-) plastic waste, it is important to include the impact of collection and sorting processes as well. In addition, these might differ for various waste treatment options.

In the next step, this framework will be applied for different case studies on a product level, in cooperation with producers. Furthermore, the concept will be extended to post-consumer waste, by including collection and sorting procedures.

Acknowledgements The authors thank hereby the German Federal Ministry of Education and Research as well as the project executing organisation within the German Aerospace Center (DLR) for the funding and support of the research project "New pathways, strategies, business and communication models for bioplastics as a building block of a sustainable economy" (BiNa) within this research has been conducted.

References

BMBF.: Destination bioeconomy—research for a biobased and sustainable economic growth. https://www.bmbf.de/pub/Destination_Bioeconomy.pdf. Assessed 01 Jan 2018

Di Maio, F., Rem, P.C.: A robust indicator for promoting circular economy through recycling. J. Environ. Prot. **6**(10), 1095–1104 (2015)

Ellen MacArthur Foundation.: Towards the circular economy: economic and business rationale for an accelerated transition. https://www.ellenmacarthurfoundation.org/assets/downloads/publications/Ellen-MacArthur-Foundation-Towards-the-Circular-Economy-vol.1.pdf (2012). Assessed 01 Jan 2018

Ellen MacArthur Foundation and Granta Design.: An approach to measuring circularity—methodology. https://www.ellenmacarthurfoundation.org/assets/downloads/insight/Circularity-Indicators_Project-Overview_May2015.pdf (2015). Assessed 01 Jan 2018

Endres, H.-J., Siebert-Raths, A.: Engineering Biopolymers—Markets, Manufacturing Properties and Applications. Hanser Verlag, München (2011)

Geissdoerfer, M., Savaget, P., Bocken, N.M.P., Hultink, E.J.: The circular economy—a new sustainability paradigm? J. Cleaner Prod. **143**, 757–768 (2017)

Geyer, R., Jambeck, J.R., Law, K.L.: Production, use, and fate of all plastics ever made. Sci. Adv. **3**(7), e1700782 (2017)

Huysman, S., De Schaepmeester, J., Ragaert, K., Dewulf, J., De Meester, S.: Performance indicators for a circular economy: a case study on post-industrial plastic waste. Res. Conserv. Recycl. **120**, 46–54 (2017)

Hildebrandt, J., Bezama, A., Thrän, D.: Cascade use indicators for selected biopolymers: are we aiming for the right solutions in the design for recycling of bio-based polymers? Waste Manage. Res. **35**(4) 367–378 (2017)

IfBB.: Biopolymers—facts and statistics. https://www.ifbb-hannover.de/files/IfBB/downloads/falt blaetter_broschueren/Biopolymers-Facts-Statistics_2017.pdf (2017). Assessed 01 Jan 2018

ISO (International Organization for Standardization): ISO International Standard 14044: Environmental Management—Life Cycle Assessment—Requirements and Guidelines. International Organization for Standardization, Geneva, Switzerland (2006)

Kirchherr, J., Reike, D., Hekkert, M.: Conceptualizing the circular economy: an analysis of 114 definitions. Res. Conserv. Recycl. **127**, 221–232 (2017)

Kreiger, M.A., Mulder, M.L., Glover, A.G., Pearce, J.M.: Life cycle analysis of distributed recycling of post-consumer high density polyethylene for 3-D printing filament. J. Cleaner Prod. **70**, 90–96 (2014)

Lieder, M., Rashid, A.: Towards circular economy implementation: a comprehensive review in context of manufacturing industry. J. Cleaner Prod. **115**, 36–51 (2016)

Linder, M., Sarasini, S., van Loon, P.: A metric for quantifying product-level circularity. J. Ind. Ecol. **21**(3), 545–558 (2017)

Ministry of the Environment—Government of Japan.: Establishing a sound material-cycle society: milestone toward a sound material-cycle society through changes in business and lifestyles. https://www.env.go.jp/en/recycle/smcs/f_plan.pdf (2003). Assessed 01 Jan 2018

NatureWorks.: Chemical Recycling. https://www.natureworksllc.com/What-is-Ingeo/Where-it-Go es/Chemical-Recycling (2017). Assessed 01 Jan 2018

PlasticsEurope.: Plastics - the Facts 2016—Any analysis of European plastics production, demand and waste data. http://www.plasticseurope.de/cust/documentrequest.aspx?DocID=67651 (2016). Assessed 01 Jan 2018

Scarlat, N., Dallemand, J.-F., Monforti-Ferrario, F., Nita, V.: The role of biomass and bioenergy in a future bioeconomy: policies and facts. Environ. Dev. **15**, 3–34 (2015)

UNEP—United Nations Environment Programme.: Assessing global resource use—a systems approach to resource efficiency and pollution reduction 2017. http://www.resourcepanel.org/fi le/904/download?token=PZsnNu_x. Assessed 01 Jan 2018

World Economic Forum.: The New Plastics Economy—Rethinking the future of plastics. http://www.weforum.org/docs/WEF_The_New_Plastics_Economy.pdf (2016). Assessed 01 Jan 2018

Spatially Differentiated Sustainability Assessment of Products

Christian Thies, Karsten Kieckhäfer, Thomas S. Spengler
and Manbir S. Sodhi

Abstract Due to the globalization of supply chains, the environmental and social impacts related to products are often dispersed over many locations. Life cycle-oriented sustainability assessment methods aim at compiling the total impacts without explicitly considering their spatial distribution. This paper illustrates how the incorporation of spatial differentiation in sustainability assessment can influence assessment results and lead to different conclusions about the design of supply chains to improve product sustainability. Comparing two alternative configurations of a simplified supply chain for beer production and concentrating only on environmental impacts, it is found that the consideration of environmental and technological heterogeneity has the potential to reverse the rank order of the alternatives.

Keywords Sustainability assessment · Spatial differentiation · Environmental heterogeneity · Technological heterogeneity

1 Introduction

While functionality, quality, and cost have typically been the predominant criteria in the assessment of products, it has been observed that sustainability aspects are now receiving increased attention (O'Rourke 2014). Product sustainability, as understood in this paper, refers to the environmental, economic, and social impacts related to the different stages of the product's life cycle. Due to the global value chains of many products, these impacts are often dispersed over multiple locations.

C. Thies (✉) · K. Kieckhäfer · T. S. Spengler
Institute of Automotive Management and Industrial Production, Technische Universität
Braunschweig, Mühlenpfordtstr. 23, 38106 Braunschweig, Germany
e-mail: ch.thies@tu-braunschweig.de

T. S. Spengler · M. S. Sodhi
Department of Mechanical, Industrial and Systems Engineering, University
of Rhode Island, 103 Gilbreth Hall, Kingston, RI 02881, USA

© Springer Nature Switzerland AG 2019
L. Schebek et al. (eds.), *Progress in Life Cycle Assessment*, Sustainable
Production, Life Cycle Engineering and Management,
https://doi.org/10.1007/978-3-319-92237-9_17

While some of the impacts, such as the emission of greenhouse gases, are of global importance, other impacts, such as the emission of acidifying or toxic substances, and particularly social impacts, are primarily applicable on a regional or even local scale.

Life cycle-oriented sustainability assessment methods (Kloepffer 2008; Finkbeiner et al. 2010; Zamagni et al. 2013), such as environmental life cycle assessment (LCA), usually compile the total impacts related to a product supply chain without explicitly considering their spatial distribution. Instead, global average values with regard to the inputs and outputs of the involved processes as well as average characterization factors are applied to compute aggregate impact indicators covering the entire product life cycle. These results are often used to identify sustainability hotspots in the product supply chain or to make comparisons between alternative products.

Neglecting the spatial dimension of the product supply chain may be problematic for the following reasons. First, *interregional trade flows* require that raw materials and products are transported over thousands of kilometers and the total transportation distance depends on the locations where the processes are carried out. Second, the inputs and outputs of manufacturing processes are influenced by *technological heterogeneity*. For example, the technologies for electricity generation and the resulting impacts are quite different throughout the world. Third, the local impacts of the same process (with identical inputs and outputs) carried out in different regions may vary due to *environmental, social, and economic heterogeneity*. As an example, the regionalized characterization factors for the acidification potential of nitrogen oxides can vary to a factor of more than 100 across European countries (Hauschild and Potting 2005). Thus, the geographic dispersion of the supply chain should not be ignored when assessing product sustainability. Instead, spatially differentiated assessments taking into account regional differences should be carried out. From this, important information about the design of the supply chain can be derived in order to improve product sustainability.

The topic of spatial differentiation has been discussed for many years in the life cycle assessment community (Potting and Hauschild 1997; Finkbeiner et al. 1998; Potting and Hauschild 2006; Reap et al. 2008; Finnveden et al. 2009). The main approaches to implement spatial differentiation within LCA are inventory regionalization, inventory spatialization, and impact regionalization (Patouillard et al. 2018). While inventory regionalization refers to the improvement of the geographic representativeness of inventory data (type and quantity of economic and elementary flows), inventory spatialization aims at attributing geographic locations to the unit processes and their corresponding elementary flows. Impact regionalization addresses the application of regionalized characterization factors that are more representative of specific geographic areas. For example, regionalized characterization factors have been developed by Hauschild and Potting (2005) and Verones et al. (2016). Computational models to perform spatially differentiated LCA calculations have been proposed by Mutel and Hellweg (2009) and Yang and Heijungs (2017).

However, in practice, spatial differentiation is rarely applied. A recent review of 142 articles in the context of operations research and sustainability assessment (Thies et al. 2018a) found that in only 20 of the articles a site-specific assessment, taking into account the local specifics, was carried out. In 20 further articles was the assessment site-dependent, considering at least some characteristics of the region or country. The most frequent use of spatially explicit data was at the inventory analysis level, but rarely with regard to the characterization factors.

Motivated by the discussion above, the objective of this paper is to show how the application of spatial differentiation in sustainability assessment can influence the assessment results and the conclusions that can be drawn on the design of supply chains in order to improve product sustainability. To this end, a hypothetical setting considering a supply chain of beer is introduced and the results for different scenarios of spatial differentiation are compared. For the sake of simplicity, the assessment concentrates on environmental impacts. Finally, the potentials and limitations of spatial differentiation in sustainability assessment related to supply chain design decisions are discussed and avenues for further research are identified.

2 Exemplary Setting for Spatially Differentiated Sustainability Assessment

In this section, a hypothetical setting to highlight the consequence of spatial differentiation in sustainability assessment is introduced. This example considers selected environmental aspects of a highly simplified supply chain for beer production and is adapted from Yang and Heijungs (2017). While the original example contains spatially differentiated data on production technologies and environmental characteristics, transportation processes and alternative supply chain configurations have been added for the purpose of our analysis.

The structure of the supply chain for beer production is depicted in Fig. 1. It comprises two production processes, grain cultivation and brewery, which are connected by transportation processes. The final product beer is to be delivered to three different regions (R1, R2, R3), which define the geographic boundaries of the supply chain. While the production processes can be carried out in each of the regions, it is assumed that the production technology (grain yields, brewing efficiency) as well as the set of environmental characteristics (sensitivity to acidifying substances) is different in each region. The functional unit of the assessment is the delivery of 1 L of beer to each of the demand regions.

The two alternative supply chain configurations under investigation are depicted in Fig. 2. In the decentralized structure, demand in each region is served by a local brewery that sources barley from local producers. In the centralized structure, demand in each region is served from a central brewery in R3, which sources all barley from a producer in R1. The transportation distances are assumed to be 50 km within a region and 200 km across regions.

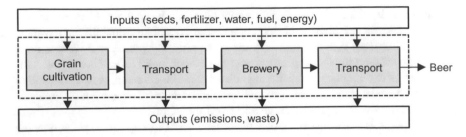

Fig. 1 Structure of the simplified supply chain for beer production

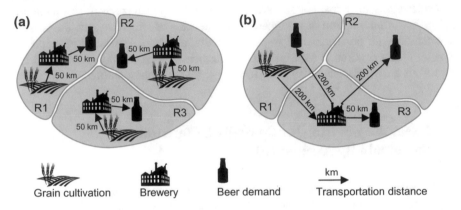

Fig. 2 Alternative supply chain configurations: (**a**) decentralized structure, (**b**) centralized structure

The process parameters for grain cultivation, brewery, and transportation are pro-vided in Table 1. Next to the global average value of each parameter, a set of spatially differentiated values is given, reflecting the *technological heterogeneity* across the regions. The yields of grain cultivation are assumed to be highest in R1, followed by R2 and R3. Consequently, the emissions per kilogram of barley are lowest in R1. The brewery technology is assumed to be identical in R1 and R2, and more efficient in R3, which is reflected by the lower barley input and lower emissions. The parameters of the transportation processes are assumed to be identical in all three regions.

Similarly, the characterization factors for impact assessment are provided in Table 2. With regard to climate change as an impact category that is effective on a global scale, the characterization factors are identical for all regions and corre-spond to the global average value. In contrast, region-specific characterization fac-tors for acidification are provided to reflect the *environmental heterogeneity* across the regions.

Table 1 Process parameters for grain cultivation, brewery, and transportation

Parameter	Unit	Global average value	Spatially differentiated values		
			R1	R2	R3
Grain cultivation	kg	0.2	0.15	0.18	0.3
CO_2 emissions per kg barley	kg	0.3	0.2	0.24	0.4
NO_x emissions per kg barley					
Brewery	kg	3	3.2	3.2	2.5
Barley input per L beer	kg	0.4	0.45	0.45	0.25
CO_2 emissions per L beer	kg	0.8	1.1	1.1	0.5
NO_x emissions per L beer					
Transportation	kg	0.7	0.7	0.7	0.7
CO_2 emissions per t*km	kg	0.4	0.4	0.4	0.4
NO_x emissions per t*km					

Table 2 Characterization factors for climate change and acidification

Parameter	Unit	Global average value	Spatially differentiated values		
			R1	R2	R3
Climate change CO_2	kg CO_2-eq	1	1	1	1
Acidification NO_x	kg H^+-eq	1.2	0.9	1.5	1.3

3 Results and Discussion

For both design options, a region-specific inventory of inputs and outputs is calculated and the impact assessment is carried out via a spreadsheet model that has been implemented in MS Excel. The results of the assessment are shown in Fig. 3 for four scenarios of spatial differentiation: (a) no spatial heterogeneity, (b) environmental heterogeneity, (c) technological heterogeneity, (d) environmental and technological heterogeneity.

In scenario (a), global average values for process parameters as well as characterization factors are applied. The total climate change and acidification impacts are higher for the centralized supply chain structure than for the decentralized supply chain structure, which is solely due to the longer transportation distances. Thus, the decentralized supply chain would be considered advantageous over the centralized structure from a global sustainability perspective.

Environmental heterogeneity is introduced in scenario (b). In this instance, the spatially differentiated characterization factors are applied, but global average values are used for process parameters. Compared to scenario (a), the results for climate change remain unchanged. However, with regard to acidification, the total impact of the centralized supply chain structure becomes lower than in the decentralized case because the emission of nitrogen oxides mainly takes place in regions where

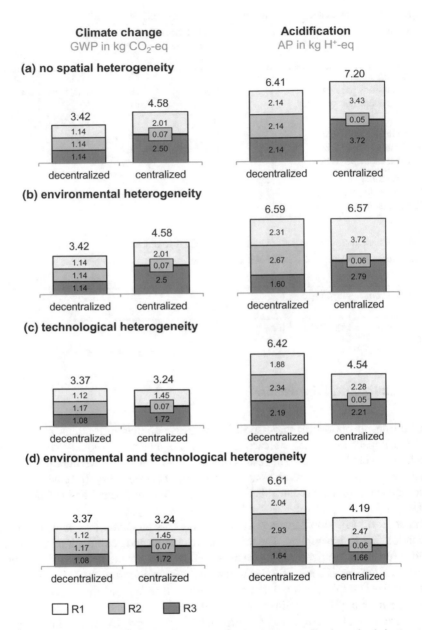

Fig. 3 Impact assessments results for the decentralized and the centralized supply chain structure in the categories climate change and acidification in different scenarios of spatial differentiation (**a–d**)

the spatially differentiated characterization factors are lower than the global average value.

Similarly, technological heterogeneity is introduced in scenario (c) by applying spatially differentiated process parameters while fixing the characterization factors to their global average values. This affects the results for both climate change and acidification. Due to the high yields of grain cultivation in R1 and the efficient brewery technology in R3, the centralized structure gains advantages over the decentralized structure in both impact categories.

Finally, environmental and technological heterogeneity are considered simultaneously in scenario (d) by applying the spatially differentiated values for process parameters and characterization factors. In this case, the effects of scenarios (b) and (c) add up, and as a result, the centralized structure has lower total impact with regard to climate change and acidification. Consequently, from a global sustainability perspective, the centralized structure should be preferred.

The results also reveal the spatial distribution of impacts. This is especially interesting for acidification as a regionally relevant impact category. It can be observed that although the total acidification impact is lower in the centralized structure, the regional acidification impact in R1 and R3 is slightly higher. Only in R2, the regional impact is much lower. Thus, a decision maker with a regional perspective on sustainability might prefer another design option than a decision maker with a global sustainability perspective.

4   Conclusions

This paper addresses the issue of spatial differentiation in sustainability assessment. Using a simplified example of beer production and concentrating on environmental sustainability aspects, it is shown that the application of region-specific process parameters and characterization factors instead of global average values influences the assessment results and has the potential to change the rank order of the alternatives. Furthermore, it gives insights into the geographic distribution of the impacts, highlighting potential conflicts of objectives between the perspectives of local and global decision makers.

The results underline the need for a consistent framework that allows for the spatially differentiated modeling of production and transportation processes and integrates regionalized impact assessment methods. Such a framework would not only enable a more accurate assessment of product sustainability, but also support the design of appropriate supply chain structures that allow for an improvement of product sustainability. Furthermore, the framework would not be limited to environmental sustainability aspects, but also integrate economic and social indicators. To this end, additional resource flows (e.g. labor) and respective characterization factors (e.g. wages, work accidents) need to be considered. A more detailed discussion on this is provided in Thies et al. (2018b).

In future work, more realistic supply chains and additional indicators for economic and social sustainability issues need to be investigated. To this end, the computational logic, which could be implemented in a spreadsheet model for the illustrative example above, needs to be formalized and implemented in a more sophisticated modeling environment. Furthermore, suitable data structures to handle the various materials and processes need to be developed.

With an increasing complexity of the supply chain, the number of possible design options to be considered than can be solved by manual iteration. The design of the supply chain should therefore be supported by appropriate optimization algorithms with options for decision makers to interact with the parameters/results and input their preferences. This would allow for the inclusion of many additional features of the decision making process, transcending the constraints set by numerical optimizations and leading to better human centered decisions.

References

Finkbeiner, M., Saur, K., Hoffmann, R., Gediga, J., Kreißig, J., Eyerer, P.: The spatial dimension in life cycle assessments. In: Total Life Cycle Conference and Exposition. SAE International (SAE technical paper series), Graz, Austria, 1–3 Dec 1998 (1998)

Finkbeiner, M., Schau, E.M., Lehmann, A., Traverso, M.: Towards life cycle sustainability assessment. Sustainability 2(10), 3309–3322 (2010). https://doi.org/10.3390/su2103309

Finnveden, G., Hauschild, M.Z., Ekvall, T., Guinée, J., Heijungs, R., Hellweg, S., Koehler, A., Pennington, D., Suh, S.: Recent developments in life cycle assessment. J. Environ. Manage. 91(1), 1–21 (2009). https://doi.org/10.1016/j.jenvman.2009.06.018

Hauschild, M., Potting, J.: Spatial differentiation in life cycle impact assessment. The EDIP2003 methodology. Danish Environmental Protection Agency (Environmental News, 80). http://www2.mst.dk/udgiv/publications/2005/87-7614-579-4/pdf/87-7614-580-8.pdf (2005). Accessed on 6 Sept 2017

Kloepffer, W.: Life cycle sustainability assessment of products. Int. J. Life Cycle Assess. 13(2), 89–95 (2008). https://doi.org/10.1065/lca2008.02.376

Mutel, C.L., Hellweg, S.: Regionalized life cycle assessment. Computational methodology and application to inventory databases. Environ. Sci. Tech. 43(15), 5797–5803 (2009). https://doi.org/10.1021/es803002j

O'Rourke, D.: The science of sustainable supply chains. Science 344(6188), 1124–1127 (2014). https://doi.org/10.1126/science.1248526

Patouillard, L., Bulle, C., Querleu, C., Maxime, D., Osset, P., Margni, M.: Critical review and practical recommendations to integrate the spatial dimension into life cycle assessment. J. Clean. Prod. 177, 398–412 (2018). https://doi.org/10.1016/j.jclepro.2017.12.192

Potting, J., Hauschild, M.: Predicted environmental impact and expected occurrence of actual environmental impact. Part II: Spatial differentiation in life-cycle assessment via the site-dependent characterisation of environmental impact from emissions. Int. J. Life Cycle Assess. 2(4), 209–216 (1997). https://doi.org/10.1007/bf02978417

Potting, J., Hauschild, M.: Spatial differentiation in life cycle impact assessment. A decade of method development to increase the environmental realism of LCIA. Int. J. Life Cycle Assess. 11(S1), 11–13 (2006). https://doi.org/10.1065/lca2006.04.005

Reap, J., Roman, F., Duncan, S., Bras, B.: A survey of unresolved problems in life cycle assessment. Part 2: Impact assessment and interpretation. Int. J. Life Cycle Assess. 13(5), 374–388 (2008). https://doi.org/10.1007/s11367-008-0009-9

Thies, C., Kieckhäfer, K., Spengler, T.S., Sodhi, M.S.: Operations research for sustainability assessment of products: A review. Eur. J. Oper. Res. (2018a). https://doi.org/10.1016/j.ejor.2018.04.03 9

Thies, C., Kieckhäfer, K., Spengler, T.S., Sodhi, M.S.: Spatially differentiated sustainability assessment for the design of global supply chains. Proc. CIRP **69**, 435–440 (2018b). https://doi.org/1 0.1016/j.procir.2017.11.001

Verones, F., Hellweg, S., Azevedo, L.B., Chaudhary, A., Cosme, N., Fantke, P., Goedkoop, M., Hauschild, M., Laurent, A., Mutel, C.L., Pfister, S., Ponsioen, T., Steinmann, Z., van Zelm, R., Vieira, M., Huijbregts, M.A.J.: LC-Impact Version 0.5. A spatially differentiated life cycle impact assessment approach. http://www.lc-impact.eu/downloads/documents/LC-Impact_report_SEPT 2016_20160927.pdf (2016). Accessed on 24 Aug 2017

Yang, Y., Heijungs, R.: A generalized computational structure for regional life-cycle assessment. Int. J. Life Cycle Assess. **22**(2), 213–221 (2017). https://doi.org/10.1007/s11367-016-1155-0

Zamagni, A., Pesonen, H.-L., Swarr, T.: From LCA to life cycle sustainability assessment: concept, practice and future directions. Int. J. Life Cycle Assess. **18**(9), 1637–1641 (2013). https://doi.or g/10.1007/s11367-013-0648-3

Life Cycle Assessment of German Energy Scenarios

Nils Thonemann and Daniel Maga

Abstract The transition of the German energy system is in full swing. Although the targets for the so-called "Energiewende" are set by the German Federal Government, different development paths are drawn by studies on energy scenarios. One major goal of the "Energiewende" is providing an electricity mix with reduced environmental impacts. Consequently, the goal of this study is to environmentally analyze the different energy scenarios. In a first step, a systematic literature review was followed to come up with 14 studies and 26 energy scenarios. After reducing the number of scenarios to five due to selection criteria, the environmental impacts of these five scenarios were analyzed applying life cycle assessment. Assumptions were made to transfer the scenarios into a sound life cycle assessment model. The life cycle assessment was conducted using the GaBi software as well as the GaBi database to investigate the environmental impacts of future electricity mixes in the years 2020, 2030, 2040, and 2050. The results show that the calculated impact on global warming of electricity generation in the different scenarios is higher compared to the greenhouse gas emissions presented in the respective studies. The differences can be explained, for example, by neglecting transmission losses in the calculation of the global warming impact in the studies.

Keywords Life Cycle Assessment · LCA · German Energy Transition · Energy Scenarios · GaBi

1 Introduction, Motivation, and Goal

The German energy transition is bringing about a huge change in the energy sector. In 2016, 31.7% of electricity was generated by renewable energies in Germany (Federal Environment Agency 2017). According to the Federal Government's goals,

N. Thonemann (✉) · D. Maga
Fraunhofer Institute for Environmental, Safety, and Energy Technology UMSICHT,
Osterfelder Straße 3, 46047 Oberhausen, Germany
e-mail: nils.thonemann@umsicht.fraunhofer.de

© Springer Nature Switzerland AG 2019
L. Schebek et al. (eds.), *Progress in Life Cycle Assessment*, Sustainable
Production, Life Cycle Engineering and Management,
https://doi.org/10.1007/978-3-319-92237-9_18

their share should increase to 80% by 2050, while reducing greenhouse gas emissions by 80–95% compared to 1990 (Federal Ministry for the Environment, Nature Conservation and Nuclear Safety (BMU) 2016; Federal Ministry of Economics and Technology (BMWi) and BMU 2010). However, in the literature, the transformation process of the energy sector towards a renewable energy-based electricity production is considered very differently (Grimmelsmann 2016).

The goal of this contribution is to select electricity scenarios and ecologically assess them. For the selection of electricity scenarios we took the results from the systematic literature review on German electricity scenarios conducted by Grimmelsmann (2016) and for the environmental analysis, a life cycle assessment (LCA) was carried out. The developed LCA models are used to ecologically evaluate future electricity scenarios which are currently applied in the Fraunhofer project "Strom als Rohstoff" and the project "Carbon2Chem" funded by the Federal Ministry of Education and Research (Federal Ministry of Research and Education [BMBF] 2016; Weidner and Pflaum 2017).

In the next section the methodological approach, which is based on a systematic literature review and an LCA, is described. Thereafter, the section contains the results and a critical discussion is given. Finally, the contribution is completed by a conclusion and an outlook.

2 Methodological Approach

A systematic literature review, conducted by Grimmelsmann (2016), has shown that there is a variety of studies that calculate future energy scenarios for Germany. In total, the review includes 14 studies from the years 2011 to 2016 including 26 energy scenarios (see Table 1).

The goal of the literature review was to identify a future energy scenario that comprise information on the development of excess energy, energy prices and related CO_2-emissions (Grimmelsmann 2016). In order to address excess energy, the future gross electricity consumption was defined as first selection criterion. In order to analyze energy prices, the investment requirements for the energy producer park were defined as second selection criterion. Finally, to estimate the related CO_2-emissions, the current expansion targets for renewable energies of the Federal Government were chosen as third selection criterion.

According to the first criterion, we choose scenarios, which are in line with the set target by the BMU (2016) to reduce the gross electricity consumption by 25% until 2050 compared to the base year 2008. Regarding the second criterion, scenarios were chosen which undercut the investment barrier of 560 billion Euros set also by the BMU (2016). For the third criterion, scenarios were chosen which go along with the expansion targets set by the German Renewable Energy Act (BMWi 2017; Grimmelsmann 2016).

Table 1 Overview of the observed studies

Author	Title	Year	Scenarios	Reference
50 HzTransmission GmbH; Amprion GmbH; TenneT TSO GmbH; TransnetBW GmbH	Network Development Plan Electricity 2025	2015	B1 B2	50 Hertz Transmission GmbH et al. (2015)
Agora Energiewende	Electricity storage in the energy transition	2014	2033 Scenario 90% Scenario	Agora Energiewende (2014)
Dena	Integration of renewable energies in the German-European electricity market	2012	Without name	Dena (2012)
DLR; Fraunhofer IWES; IfnE	Long-term scenarios and strategies for the expansion of renewable energies in Germany, taking into account developments in Europe and globally	2012	A THG 95	DLR et al. (2012)
Fraunhofer ISE	Energy system Germany 2050	2013	DE 2050	Fraunhofer ISE (2013)
Fraunhofer IWES	Analysis of Power-to-Gas energy storages in the renewable energy system	2014	Basis	Fraunhofer IWES (2014)
IAEW; consentec	Evaluation of flexibilities of electricity generation and CHP plants	2011	Basis scenario	IAEW and consentec (2011)
IAEW; Fraunhofer IWES	Roadmap storage	2014	C	IAEW and Fraunhofer IWES (2014)
Nitsch, Joachim	Current scenarios of the German energy supply taking into account the key figures of the year 2014	2015	Corridor Corridor-Over Corridor-Under Scenario 100	Nitsch (2015)
Nitsch, Joachim	The Energiewende after COP 21—Current Scenarios of German Energy Supply	2016	Trend Climate 2040 Climate 2050	Nitsch (2016)
Öko-Institute; Fraunhofer ISI	Climate protection scenario 2050—2. Final report	2015	AMS KS 80 KS 95	Öko-Institute and Fraunhofer ISI (2015)
Prognos	Development of Energy Markets - Energy Reference Forecast	2014	Trend scenario Target scenario	Prognos (2014)
UBA	Projection report 2015	2015	MMS	UBA (2015)
VDE ETG	Energy storages in the Energiewende	2012	40% Scenario 80% Scenario	VDE ETG (2012)

After applying the selection criteria, the number of scenarios was reduced to five. The "Climate Protection Scenario 80 (KS80)", the "Current Measures Scenario (AMS)" and the "Climate Change Scenario 95 (KS95)" from Öko-Institute and Fraunhofer ISI (2015), as well as Prognos (2014) "Trend Scenario (Trend)" and "Target Scenario (Target)" were selected.

To calculate the environmental impacts of the different electricity mixes the LCA software GaBi was used. As scope we use the system boundary starting from the extraction of resources to the provision of the electricity to the end consumer. The functional unit was chosen to be the provision of 1 kWh. Since the main political goal of the energy transition is to reduce greenhouse gas emissions, the LCA study focuses on climate change. The following assumptions were made to transfer the information contained in the scenario descriptions to the requirements of the LCA model.

Pumped storage power plants, for example, were considered to operate with an overall efficiency of 80% (Giesecke et al. 2014). The required electricity for pumping is expected to be provided by the national electricity mix. Future import surpluses are considered with the EU-28 power mix of the respective year depicted by processes of the GaBi Database: Extension II: energy 2017 (Thinkstep 2017). However, export of electricity to other European countries is neglected since there is no disaggregated data for exports by energy sources in the given scenarios. Additionally, the export does not influence the environmental impacts of the German electricity mix. This can be explained by the fact that, if export is taken into account, the relative composition of the net electricity providers is not changed. Imports or exports via high-voltage direct current cable is not considered in the analysis as it has only a small total contribution and almost identical amounts of electricity are imported or exported (Öko-Institute and Fraunhofer ISI 2015). For transmission losses, a value of 6.37% was calculated according to Federal Statistical Office (Destatis) (2017).

Besides these general assumptions, special assumptions for the LCA modeling of the respective studies were made. For the energy scenarios from Öko-Institute and Fraunhofer ISI (2015), the net electricity providers summarized under the term "others" in the sub-category "other" were allocated to the remaining net electricity producing technologies. For the AMS, the electricity generation by municipal waste incineration plants was concluded to remain constant over time and was therefore set to the fixed value of the reference year 2012 (6.9 TWh). According to the KS80 and the KS95, 0.6 TWh of annual electricity production were assigned to "other" net electricity production sources in the sub-category "other". Heavy oil and coal gases were distributed on the basis of the partitioning in the climate protection scenario 80 and the climate protection scenario 95.

Special assumptions were also made for the Prognos (2014) scenarios. Similarly to the aforementioned scenarios, the Prognos (2014) scenarios provided a little-detailed breakdown of net electricity providers, hence, assumptions were made on the same basis as for the other scenarios introduced above. The net electricity providers' solid biomass and biogas were differentiated according to the primary energy consumption for the respective sectors (heat generation, power generation, and mobility) mentioned in Prognos (2014). The net electricity provider category "other" was distinguished in geothermal energy, mine gas, sewage and landfill gas, as well as in biogenic and other wastes. For the provision of electricity from geothermal sources a rise to 0.7 TWh per year by 2020 according to the Federal Republic of Germany (2012) was assumed and that thereafter the contribution would remain at a relatively constant level (Prognos 2014). The electricity generation via mine gas, sewage gas, and landfill gas is supposed to decrease over time (Prognos 2014). Energy from waste incineration, on the other hand, was expected to increase slightly to about 6 TWh due to the increased incineration of waste (Prognos 2014).

The choice of the net electricity producing technologies was one of the crucial parts of this LCA study. For consistency reasons, only processes from the GaBi database were used. Net electricity producing technologies were assigned to the LCA processes as shown in Table 2. In Table 3 the share of energy sources for KS95 over different years is listed. The mentioned assumptions were applied to the life cycle inventory for the electricity mixes in GaBi.

3 Results and Discussion

In this section, the results are presented and discussed. The results focus on the global warming potential over a time horizon of 100 years (GWP_{100}). First, the calculated global warming potential (GWP) of the energy scenarios will be introduced. Secondly, the contribution of different energy sources to the overall GWP will be shown. Thirdly, the differences of the calculated GWP values to the GWP values of the energy mixes from GaBi will be presented. Finally, the dependency of energy-intensive processes on the selection of background data for the energy mix will be investigated as an example.

Figure 1 shows the GWP of the different scenarios as calculated via LCA and the values for GWP as stated in the reports on the electricity mix scenarios. It can be observed that in all scenarios the GWP is decreasing over time, which corresponds to the target set by BMWi and BMU (2010). However, there are discrepancies between the calculated values and the values, which are given in the respective studies.

For a more detailed analysis, the relative deviation between the calculated GWP values and those taken from the studies is shown in Table 4. The results reveal that the discrepancies between the calculations and the study results differ depended on the selected scenario and year. The AMS and Trend scenarios are reflected best in terms of relative discrepancy. Nevertheless, especially the calculated GWP for KS95 demonstrates a large relative difference; for example, in the year 2040, the relative discrepancy is 205.17%. This means, for this case, the calculation gives a value approximately two times larger than the value reported in the study. Consequently the contribution of processes is investigated for KS95 in the following.

Focusing on KS95, in Fig. 2 the relative contribution of different energy sources to the GWP is shown. For the energy mixes for 2012–2040 the energy carriers 'lignite and hard coal are contributing the most to the GWP. In 2050, the GWP is resulting from various energy sources, but mostly by energy produced from natural gas,

Table 2 Net electricity
producing technologies
assigned to processes from
the GaBi database

Processes from GaBi database	Net electricity producing technology	
	Öko-Institute and Fraunhofer ISI (2015)	Prognos (2014)
DE: Electricity from nuclear ts	Nuclear energy	Nuclear energy
DE: Electricity from lignite ts	Lignite	Lignite
DE: Electricity from hard coal ts	Hard coal	Hard coal
DE: Electricity from coal gases ts	Refinery gas	–
	Blast furnace gas	
	Coke oven and town gas	
DE: Electricity from natural gas ts	Natural gas	Gas
	Backup-plants	
DE: Electricity from heavy fuel oil (HFO) ts	Oil	Fuel oil
DE: Electricity from biomass (solid) ts	Biomass	–
	Vegetable oil	
DE: Electricity from biogas ts	Biogas	–
DE: Electricity from waste ts	Waste	–
DE: Strom aus Windkraft	Wind onshore	Wind power onshore
	Wind offshore	Wind power offshore
DE: Electricity from photovoltaic ts	Solar	PV
DE: Electricity from hydro power ts	Water	Run-of-river power plants and storage water
IT: Electricity from geothermal ts	Other renewable energy	–

Table 3 Share of energy sources within KS95

Energy source	2012 (%)	2020 (%)	2030 (%)	2040 (%)	2050 (%)
Lignite	25.24	14.59	1.03	1.57	0.00
Hard coal	18.20	11.13	6.26	2.98	0.00
Coal gases	1.55	1.78	1.30	0.43	0.01
Nuclear	16.06	12.61	0.00	0.00	0.00
Natural gas	12.80	12.55	19.25	11.24	2.32
Heavy fuel	1.06	0.58	0.10	0.02	0.00
Wind onshore	8.49	19.96	31.68	38.81	51.02
Wind offshore	0.12	5.17	10.50	18.30	23.56
Biogas	4.48	4.12	2.59	0.55	0.08
Biomass	2.62	2.54	2.39	0.82	0.42
Hydro power	3.70	4.38	4.76	3.68	3.23
Photovoltaic	4.50	9.07	13.61	12.31	16.15
Waste	1.18	1.22	1.09	0.67	0.47
Geothermal	0.00	0.22	0.86	1.24	1.61
Import	0.00	0.06	4.59	7.38	1.14

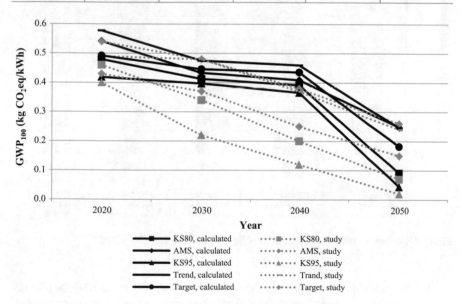

Fig. 1 GWP_{100} of the calculated and given results of the energy scenarios over time

photovoltaics and wind onshore. The deviation of 205.17% between the calculated GWP values based on KS95 and those presented by KS95 for the year 2040 can be explained by the underestimation of the impact of the electricity production from lignite and hard coal on global warming within KS95.

Table 4 Relative
discrepancy between the
calculations and the results
from the studies

	KS80 (%)	AMS (%)	KS95 (%)	Trend (%)	Target (%)
2020					
	4.24	0.32	4.77	17.91	14.45
2030					
	21.43	−9.68	80.50	−1.14	20.49
2040					
	96.02	7.71	205.17	24.42	74.52
2050					
	30.13	−4.60	121.03	2.38	21.46

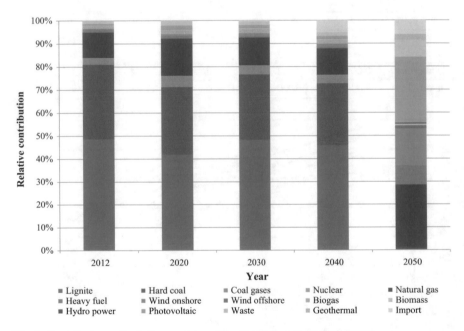

Fig. 2 Contribution analysis for processes of the electricity mixes within KS95

In order to analyze the impact of choosing background data of electricity production for an LCA of energy-intensive products, the highly energy-consuming production process of hydrogen via electrolysis was selected. According to Carmo et al. (2013) the electricity demand for the production of one Nm^3 hydrogen (H_2) via polymer electrolyte membrane (PEM) electrolysis is 5 kWh. The results of this sensitivity analysis, where the lowest and highest values of the GWP for electricity production considering the calculated values were chosen, are shown in Fig. 3.

The difference between the scenarios varies between 0.08 kg CO_2 eq./Nm^3 H_2 to 1.01 kg CO_2 eq./Nm^3 H_2. Relatively the range lies between 3 and 460% and

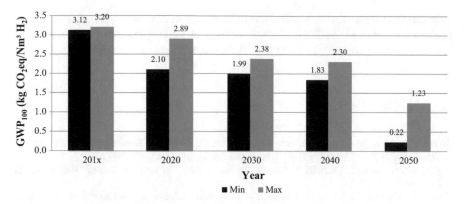

Fig. 3 GWP$_{100}$ for the electricity production of hydrogen electrolysis over time

thereby the relevance of the selection of the different energy scenarios is proven. For consideration of these differences, sensitivity analysis should be conducted to show the range of total impacts especially for product systems which are mainly influence by electrical processes.

4   Conclusion and Outlook

LCA results are often sensitive to the impact of the electricity mix. Consequently, in many cases, the selection of background data for the electricity mix is very relevant to the final outcome of an LCA. It was shown that deviations exist between the conducted GWP calculations of the energy scenarios and the results presented in the observed studies. One reason for the observed discrepancies between the calculated values for the GWP and the reported GWP values in the studies is, that different assumptions to transfer the energy scenarios into LCA models were taken into account. One assumption, for example, is that electricity imports in future energy scenarios were neglected for calculating the GWP within the studies. Another reason is the usage of different background data in the studies as in this LCA model GaBi-datasets were used as background data. For example, in the case of electricity production from lignite and hard coal within KS95, the differing background data caused the underestimation of the GWP.

As an outlook, further research is needed to investigate and analyze the impact of future energy scenarios within additional impact categories. Results are already available but have not yet been analyzed. It is also important to consider that current LCA models work with background data representing the present state of the art technology of power generation systems. The compliance with future limit values for emissions and the environmental impact of future energy carriers are only partly taking into account. The consideration of technological developments of power generation systems for estimating the impact of future energy scenarios is not yet adequately reflected. Additionally, future research should focus on the deviations of the LCA results and how to build a LCA model with smaller uncertainties.

References

50 Hertz Transmission GmbH, Amprion GmbH, TenneT TSO GmbH, TransnetBW GmbH: Network Development Plan Electricity 2025. [Netzentwicklungsplan Strom 2025] (2015)

Agora Energiewende: Electricity Storage in the Energy Transition. [Stromspeicher in der Energiewende] (2014)

Carmo, M., Fritz, D.L., Mergel, J., Stolten, D.: A comprehensive review on PEM water electrolysis. Int. J. Hydrogen Energy **38**(12), 4901–4934 (2013). https://doi.org/10.1016/j.ijhydene.2013.01. 151

Dena: Integration of Renewable Energies in the German-European Electricity Market. [Integration der erneuerbaren Energien in den deutschen-europäischen Strommarkt] (2012)

DLR, Fraunhofer IWES, IfnE: Long-term Scenarios and Strategies for the Expansion of Renewable Energies in Germany, Taking into Account Developments in Europe and Globally. [Langfrist-szenarien und Strategien für den Ausbau der erneuerbaren Energien in Deutschland bei Berücksichtigung der Entwicklung in Europa und global] (2012)

Federal Environment Agency: Renewable Energies in Germany - Data on the Development in 2016 [Erneuerbare Energien in Deutschland - Daten zur Entwicklung im Jahr 2016]. Dessau-Roßlau (2017)

Federal Ministry for the Environment, Nature Conservation and Nuclear Safety (BMU): Climate Protection Plan 2050 - Climate Policy Principles and Objectives of the Federal Government [Klimaschutzplan 2050 - Klimaschutzpolitische Grundsätze und Ziele der Bundesregierung]. Berlin (2016)

Federal Ministry of Economics and Technology (BMWi): Renewable Energy Sources Act. EEG 2017 (2017)

Federal Ministry of Economics and Technology (BMWi), Federal Ministry for the Environment, Nature Conservation and Nuclear Safety (BMU): Energy Concept for an Environmentally Friendly, Reliable and Affordable Energy Supply [Energiekonzept für eine umwelt schonende, zuverlässige und bezahlbare Energieversorgung]. Berlin (2010)

Federal Ministry of Research and Education (BMBF): Carbon2Chem (2016)

Federal Republic of Germany: National Renewable Energy Action Plan in accordance with Directive 2009/28/EC on the promotion of the use of energy from renewable sources (2012)

Federal Statistical Office (Destatis): Production - Monthly Report on the Supply of Electricity [Erzeugung - Monatsbericht über die Elektizitätsversorgung]. https://www.destatis.de/DE/Zahle nFakten/Wirtschaftsbereiche/Energie/Erzeugung/Tabellen/BilanzElektrizitaetsversorgung.html (2017). Accessed 31 May 2017

Fraunhofer ISE: Energy System Germany 2050. [Energiesystem Deutschland 2050] (2013)

Fraunhofer IWES: Analysis of Power-to-Gas Energy Storages in the Renewable Energy System. [Analyse von Power-to-Gas-Energiespeichern im regenerativen Energiesystem] (2014)

Giesecke, J., Heimerl, S., Mosonyi, E.: Hydropower Plants - Planning, Construction and Operation [Wasserkraftanlagen - Planung, Bau und Betrieb], 6th edn. Springer, Berlin (2014)

Grimmelsmann, M.: Identification of future scenarios for the electrical energy supply system in Germany - Master thesis in cooperation with Fraunhofer UMSICHT and University Duisburg-Essen. [Identifizierung von Zukunftsszenarien für das elektrische Energieversorgungssystem in Deutschland - Masterarbeit in Kooperation zwischen Fraunhofer UMSICHT und der Universität Duisburg-Essen]. Universität Duisburg-Essen (2016)

IAEW, consentec: Evaluation of Flexibilities of Electricity Generation and CHP Plants. [Bewertung der Flexibilitäten von Stromerzeugungs- und KWK-Anlagen] (2011)
IAEW, Fraunhofer IWES: Roadmap Storage. [Roadmap Speicher] (2014)
Nitsch, J.: Current Scenarios of the German Energy Supply Taking into Account the Key Figures of the Year 2014. [Aktuelle Szenarien der deutschen Energieversorgung unter Berücksichtigung der Eckdaten des Jahres 2014] (2015)
Nitsch, J.: The Energiewende after COP 21—Current Scenarios of German Energy Supply. [Die Energiewende nach COP 21 - Aktuelle Szenarien der deutschen Energieversorgung] (2016)
Öko-Institute, Fraunhofer ISI: Climate Protection Scenario 2050—2. Final Report [Klimaschutzszenario 2050 - 2. Endbericht]. Berlin, Karlsruhe (2015)
Prognos: Development of Energy Markets—Energy Reference Forecast. [Entwicklung der Energiemärkte - Energiereferenzprognose]. Basel (2014)
Thinkstep: GaBi Software-Systema and Database for Life Cycle Engineering. Extension Database II: Energy 2017. http://www.gabi-software.com/deutsch/index/ (2017). Accessed 29 Apr 2016
UBA: Projection Report 2015. [Projektionsbericht 2015] (2015)
VDE ETG: Energy Storages in the Energiewende. [Energiespeicher für die Energiewende] (2012)
Weidner, E., Pflaum, H.: Lead project "Electricity as raw material" - Efficient electrochemistry for sustainable chemical products [Leitprojekt „Strom als Rohstoff" - Effiziente Elektrochemie für nachhaltige Chemieprodukte]. In: Neugebauer, R. (ed.) Resource Efficiency - Key Technologies for Business & Society [Ressourceneffizienz - Schlüsseltechnologien für Wirtschaft & Gesellschaft], 1st edn, pp. 197–238. Springer, Berlin (2017)

Printed in the United States
By Bookmasters